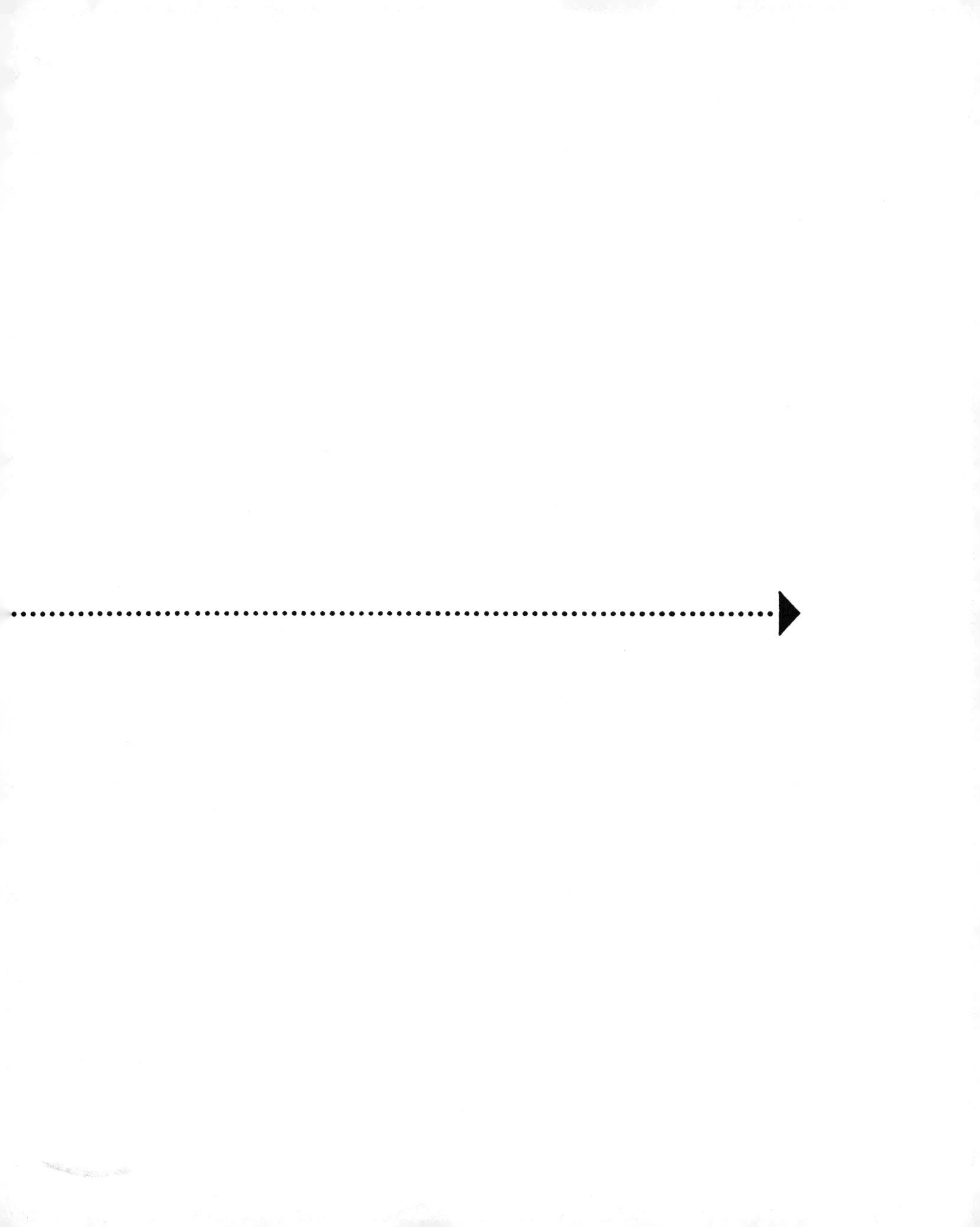

Never Leave Well Enough Alone

By

raymond loewy

with a new introduction by

glenn porter

THE JOHNS HOPKINS UNIVERSITY PRESS
BALTIMORE AND LONDON

Printed in the United States of America on acid-free paper

Hardcover originally published in 1951 by Simon and Schuster, New York
Johns Hopkins Paperbacks edition, 2002
9 8 7 6 5 4 3 2 1

The Johns Hopkins University Press
2715 North Charles Street
Baltimore, Maryland 21218-4363
www.press.jhu.edu

The original book was designed by the author, his first experiment in the field of book manufacturing. The types are Electra for the text and, occasionally, Bank Script for the headings. The format is 6 x 7 inches. The wider outer margins of each page are meant to afford better thumb space.

Library of Congress Cataloging-in-Publication Data

Loewy, Raymond, 1893–1986.
Never leave well enough alone / by Raymond Loewy with a new introduction by Glenn Porter.
p. cm.
Originally published in 1951 by Simon and Schuster, New York.
ISBN 0-8018-7211-1 (alk. paper)
1. Loewy, Raymond, 1893–1986. 2. Industrial designers—Biography. I. Title.

TS140.L63 A36 2002
745.2'092—dc21 2002069397
[B]

A catalog record for this book is available from the British Library.

"ONE MUST BE A WORK OF ART, OR WEAR A WORK OF ART."—WILDE

It has been my good fortune, in life, to do the things I have felt best equipped to do. I enjoy the process very much, and besides, it requires less effort. My assistants, to whom I owe a major part of my relative success, have been selected on the same basis of "fitness to function." It is a pleasure to watch them do things well and effortlessly.

Now, for a change, I will open myself to their criticism, by trying to do a thing for which my inadequacy is utter: that is, the writing of a book in the native tongue without benefit of professional assistance.

I feel very self-conscious about it and somewhat terrified. Mine must be the sort of feeling experienced by the trapeze artist about to perform his first aerial somersault without a net.

For the reader with masochistic tendencies who might endeavor to detect the spring from which such a prose can emerge, here are some clues: I first studied English at the Chaptal College in Paris, for a period of six months, with a paraplegic German teacher who hated anything remotely Anglo-Saxon. Next I spent some time with the American forces during World War I, but the units I worked with were mostly from Brooklyn (N. Y.). So I landed in America with practically no knowledge of the language. Whatever I know now, I have picked up since, as I have gone along trying to make a living. The "struggle for life" was nothing compared with trying to master the pronunciation of TH.

Among the factors that shaped up my linguistic stockpile are some definite influences. Most marked are Oscar Wilde, Mammy Yokum, and, lately, Edith Sitwell. Others were Amos 'n Andy, Damon Runyon, Shakespeare, and Dick Tracy. The Gimbels ads are mostly responsible for my basic (or should it be "basement"?) English. Also a person named Solomon Petofsky who ran a delicatessen on Third Avenue at Sixty-seventh Street when I first arrived in New York.

With this literary background in mind, the reader will understand that any resemblance between a living person and whatever character I will struggle to depict is intentional but purely accidental. May I repeat that this book has not been ghost-written; a statement that is both a tribute to veracity and an expression of homage to the literary pride of the Ghost-Writers Guild.

To my readers who shudder at the thought of what is going to happen to the English language between the covers of this book, may I point out in self-defense that the language hasn't treated me too gently either. After thirty years in America, South Bend to me still refuses to be anything but Souse Bend. Yet, readers will not sail in uncharted waters. Astern and to port, in front and at left, he will meet familiar landmarks: clichés, clichés of all kinds. Clichés to native Americans, but very American to a late citizen.

To those about to witness the slaughter, may I offer a limited measure of sympathy and beg relative forgiveness. And may God bestow His mercy upon the thousands of innocent infinitives now on the verge of being split.

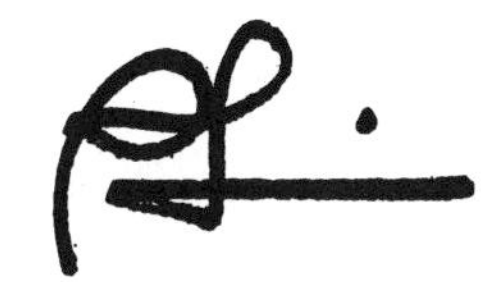

Introduction

"As soon as I form a conception of a material or corporeal substance, I simultaneously feel the necessity of conceiving that it has boundaries of some shape or other; that relatively to others it is great or small; that it is in this or that place, in this or that time, that it is in motion or at rest; that it touches, or does not touch, another body; that it is unique, rare or common; nor can I, by any act of imagination, disjoin it from these qualities."

Having thus described, five centuries ago, his approach to three-dimensional perception, Galileo established the philosophy of industrial design. He also established the fact that the earth was spherical—for which he was tortured and destroyed. Many will think that Galileo's inquisitors got all mixed up and that he should have been paid instead for his contribution to the industrial design professon.

Regardless, his lucid statement stands, and if it is good enough for G. it is good enough for me.

As to the purpose of this book, it is only fair to state now that it is not at all intended to be a textbook or to be a treatise on technology and civilization, a subject that has been treated with extraordinary brilliance and ivory tower obtuseness by many pedantic theorists. This book is the story of a young man who came to America to make a living, and simply happened to do so in a profession which he helped to create. It does not pretend to be anything else.

Acknowledgements

My indebtedness to the Atlantic Ocean must be acknowledged first. In perverse conspiration with the S.S. Nieuw Amsterdam, it lengthened my crossing to Europe and compelled me to remain in bed in my stateroom for the greater part of the journey. To this forced immobilization can be attributed the first sixty-two pages of this thing. For continued encouragement, I am deeply grateful to G. D. Searle and Company, makers of Dramamin, a new seasickness remedy.

Thanks are in order to the Pullman Company, whose new type of automatic folding toilets makes me appreciate the joys of staying home. To home, where a defective incinerator poisons my daily life, thanks for making me appreciate the joys of travel. For continual criticism of what I was thinking, doing, planning to do or write, thanks to my beloved wife, Viola, without whom this book might otherwise have been much longer. Selection of the text was greatly facilitated by my secretary, Miss Peters, whose well-timed loss of a particularly boring chapter in a New York taxicab led to its complete elimination, and my blissful relief. To the mosquitoes who made writing unbearable on the beach at Porquerolles, and chased me to Zermatt, I must credit a lovely month of June in the shadow of the Matterhorn.

To the makers of my ball point pens, may I extend the thanks of the dry cleaning industry, which has been kept busy removing spots from most of my bedsheets, pajamas, tablecloth, evening shirts, white poodles, upholstery, and Lanvin neckties during the genesis of this book. To the airlines, thanks are in order for the countless hours of leisurely waiting at airports and bus terminals, where many chapters *have been written on empty popcorn boxes, travel folders to Mexico, and other deadly airline literature. Acknowledgements are in order to Ella, my cook, who reduced the printing cost of this book by conveniently dropping a saucerful of hollandaise on a batch of illustrations, thereby materially cutting down printing expenses.*

Finally, my heartiest thanks to my dear friend Peggy (Mrs. Howard) Cullman, who, after reading the first two parts of the ms., assured me that she had read much worse, thereby supplying the final dose of enthusiasm that I so badly needed to finish the job.

PART THREE

Introduction
to the Johns Hopkins Edition

The influential and intriguing Raymond Loewy (1893–1986) was the most prominent of the founders of a new profession—industrial design—which arose around 1930 and had a profound impact on the United States and then on much of the rest of the globe. As *Cosmopolitan* magazine marveled shortly before the 1951 publication of his autobiography, *Never Leave Well Enough Alone*, "Loewy has probably affected the daily life of more Americans than any other man of his time." Designers such as Loewy, Walter Dorwin Teague, and Henry Dreyfuss changed the look and the feel of our material world. In the nation's bicentennial year a writer for the *New York Times* recalled that "when Raymond Loewy and the century were young, the look of life was different. Everyday objects were bulky, colors were dowdy and machines bristled with ungainly protuberances." Then the dark interiors and ungainly machines of the century's early years gave way in the 1920s and 1930s to the brighter, smoother, more colorful, and constantly changing look of the modern era. Americans welcomed this designed world: the modern age meant endless novelty in goods and services, in the interiors of homes, stores, and offices, in automobiles and appliances, and in all the things of the manufactured world.

The coming of mass production had removed ancient constraints on output and unleashed enormous quantities of goods. But soon this progress brought unanticipated problems—saturated markets and the challenge of ensuring sufficient consumption to sustain the constant growth upon which the society depended. The rise of modern advertising early in the twentieth century provided the first and most powerful tool for engineering consumption, but producers soon realized that turnover and sales could be further increased if the colors, shapes, and styles of goods changed more frequently. Although that was not the sole mission of industrial design, it was the primary impetus for its rise as a profession. Designers became vital cogs in the machinery of consumer capitalism and the abundant material life of the era. Looking back on the five decades of industrial design since its beginnings in the latter 1920s, Loewy reflected that he and his "early colleagues" had "helped create the life style of Americans and, by osmosis, of the rest of the world." Industrial design played a key role in the realization of the American Dream and the consumer culture that fueled the economy and gave meaning to modern life.

Raymond Loewy ran the largest and most powerful of the nation's consultant design organizations. By the time it reached its zenith at midcentury, the Loewy operation employed more than two hundred people and had offices in New York, Chicago, London, Paris, and several other cities in the Americas. He and his colleagues designed an almost incomprehensibly wide range of objects and interiors, items large and small in almost every category of industry from lipsticks to locomotives, as Loewy put it. Their automobiles, locomotives and passenger trains, buses, ocean liners, airplanes, appliances, and countless other products, packages, trademarks, and installations at world's fairs, department stores, offices, and shopping centers gave a quickened pace and a modern feel to the new age. Loewy's prominence in transportation from the 1930s into the 1950s made him the best-known figure in the predominant style of that time—streamlining. His firm worked for many years for the nation's leading corporations—the Pennsylvania Railroad, Frigidaire, Nabisco, Greyhound, Studebaker, Shell, International Harvester, Coca-Cola, and scores of others. The Loewy studios crafted such icons as the Lucky Strike cigarette packet, the Exxon name and logo, the livery and decorations for Air Force One, the Sears Coldspot refrigerator, and the habitats for many of the manned vehicles of the National Aeronautics and Space Administration. The Loewy operation also became the largest store planning and retail design service in the world in the 1950s and 1960s. Its many-sided presence extended to the globe's leading economies in the era following World War II as Loewy transplanted the American form of industrial design to Western Europe and to the developing nations. Even the Soviet Union turned to Raymond Loewy in the 1970s in hopes of having his magic improve its exports and invigorate its struggling economic system. Few persons stood so prominently as Raymond Loewy in the economic and cultural changes that shaped the twentieth century.

Although he affected American life so deeply, Loewy was born in Paris and spent his first twenty-five years in his native France. The child of a bourgeois Parisian family, he had long shown an interest in engineering as well as art, aesthetic matters, and living well, and he had begun an engineering education when World War I intervened. Loewy served with distinction in that conflict and emerged with several decorations and a captain's rank. Following the deaths of both his parents in the worldwide influenza epidemic of 1918–19, his two older brothers immigrated to the United States. In the autumn of 1919 Raymond followed them, sailing on the *France* for New York with notions of working in the electrical industry.

On the voyage Loewy happened to contribute a sketch of a fashionably dressed female passenger to a shipboard charity auction, and it fetched a good sum. The winner was the British Consul in New York, who urged Loewy to pursue a career in commercial illustration. Once in Manhattan, the young immigrant decided to follow this advice. Soon his drawings appeared in advertisements in leading magazines of the time, including *Vogue, Harper's Bazar, Vanity Fair, The New Yorker,* and *National Geographic.* He earned a good living in commercial art, usually working on ads for leading department stores, especially Saks and Bonwit Teller, depicting willowy women in fashionable gowns, hose, gloves, silk underwear, and shoes. He often placed sleek automobiles and cruise ships in these vignettes of upper-middle-class life, and soon those machines attracted favorable comment.

As he tells the story in his autobiography, Loewy eventually tired of this career in advertising illustration and longed to do something closer to his early interest in engineering. American products, he felt, were marvels of production and functionality but were also unnecessarily and unbearably ugly, noisy, smelly, and offensive. He proposed to join a "great crusade" to improve them and thus better the lives of ordinary folk. Near the end of the 1920s he entered the emerging ranks of industrial designers.

Although many so-called art industries, such as jewelry, pottery and glassware, furniture, and the like, had long used specialized designers to create their changing patterns and styles, this was not true for most mass-produced goods. But once Alfred Sloan led General Motors and the automobile industry into the revolutionary new era of competition through appearance, with an array of colors and annual model changes, styling came to many more sectors of the economy. Whenever markets became saturated, or the economy sagged, or the pressures of competition dictated, generalists in industrial design proved capable of improving the sales of American businesses. As one of the first practitioners, Loewy had the talent, drive, wit, and charisma that quickly carried him to the to the forefront of his chosen field.

This autobiography recounts Loewy's youthful days, his move to America, his early career in illustration for advertising, and his spectacular success in industrial design. When *Never Leave Well Enough Alone* appeared in 1951, Loewy was near the height of his achievements. In 1949 he had been the subject of a long piece in

Life, and in October of that year he had appeared on the cover of *Time*, the first designer ever to do so. This was one of the great distinctions of American life, and it symbolized Loewy's celebrity status. He had a genius for promoting himself and his profession, and he appeared countless times in magazines, in newspapers, and on radio and television programs both in the United States and Europe. He became a celebrated figure in the popular culture of mid-twentieth-century America, as his designs, his appealing personality, and his lavish lifestyle attracted widespread attention. He lived well, with multiple glamorous homes in North America and Europe, a changing array of yachts and customized personal cars, a dandy's wardrobe, and a penchant for partying with the rich, the famous, and the beautiful. He was always ready with surprising predictions about the future and attention-grabbing suggestions for making modern life more comfortable and more efficient. Loewy became a symbol of progress, of the good life, of all things up-to-date and desirable.

Never Leave Well Enough Alone found success both in the United States and abroad. Loewy's work made him known around the world, and translations of his book appeared in French, German, Dutch, Japanese, Spanish, and Arabic, usually under a title closer to *Ugliness Sells Poorly* than to the more complex English title. It was particularly a hit in Germany. Reviewers praised both its insights into the process of industrial design and its reflection of the many-faceted personality of one of the most fascinating and lively men of the day. Some criticized its all too clear revelation of the author's oversized ego. *The New Yorker* sniffed that it was "not always in the best of taste" and described it as "instructive, brash, cocksure, occasionally funny, sometimes vulgar, and always honest." The book is a highly personal document, even by the standards of autobiographies, and the author's persona and idiosyncracies are apparent from the first page. Loewy had strong opinions on almost everything, and he showcased his sensibilities, his wit and wisdom, even his feelings about American food (bland) and some of his favorite recipes. He asserted that this work was his own, and this claim rings true. Loewy wrote well, and though he no doubt made use of his extraordinarily talented public relations staffer Elizabeth Reese in fashioning and in promoting the autobiography, it speaks in his voice. He also designed the original book, chose the many typefaces and illustrations, and created the cover and jacket. This volume stood as a tour de force of ego, pure Loewy.

Never Leave Well Enough Alone provides a striking picture of the complex work of a nascent profession, more from the point of view of a chief executive officer and a salesman than of a journeyman designer. It was Loewy's task to travel endlessly, trying to persuade skeptical or hostile business executives and production engineers that his firm could help them make better, more appealing products. Once he established his business, Loewy, like the other leaders of large consultant design firms, provided oversight and a sense of direction but delegated most of the day-to-day design tasks to subordinates. All the firm's designs were said to be his, and his name went on all the drawings because the wider his fame, the easier it was to get work for the business. Loewy's primary job was to be the firm's brand, to win and then to retain clients, and to hire and manage talented staffers. He often said that industrial design was 20 percent inspiration and 80 percent transportation, and he recalls in the autobiography countless dreary "business-getting" trips on the road in the industrial heartland of the Midwest, with bad food, grim hotels filled with loutish conventioneers, days that finally finished at the end of a "long corridor . . . past a thousand doors, transoms open, blaring radio broadcasts, drunken arguments, riotous laughter, girlish giggling, and so often, a woman's sob." It was not always champagne and Palm Springs, especially in the early years. Indeed, Loewy's story paints almost as vivid a picture of the gray world of American business in the middle decades of the twentieth century as Arthur Miller's *Death of a Salesman.* But Raymond Loewy was no Willy Loman.

Loewy's achievements made him a man of influence, wealth, and fame in what the skeptical *New Yorker* reviewer termed the "new and sometimes appalling art form" of industrial design. When cartoonist Rube Goldberg (whose drawings of outlandish machines appear in this book) introduced Loewy at a dinner not long before the book appeared, he quipped that "an industrial designer is a person who talks another person into changing the shape of something that looks pretty good as it is." From the first, the field suffered from this lingering accusation of superficiality, of being altogether too similar to advertising. It did have much in common with that huckstering profession, and many of its top figures, such as Loewy and Walter Dorwin Teague, had backgrounds in advertising and close ties to leading agencies. But industrial design extended well beyond merchandising, providing a unique link between marketing and production. It took into account not only appearance but also efficiency of manufacturing, quality and durability, new materials, ease and safety of use, and other factors. Later its concerns extended to the environmental impact of designs and how to make products usable by the disabled. When practiced well, industrial design made genuine contributions to a better life for consumers.

Loewy's own designs sought those broader goals, and they combined engineering and artistic sensibilities. All his life he advocated simplicity of design, which he defined as reduction to essentials, the engineer's commitment to minimizing the use of materials and combining parts. A good designer, in Loewy's view, should also smooth surfaces by removing or concealing protuberances, hinges, and other unsightly mechanical items such as motors, rods, wires, and the like. This would normally result in a simple, functional, modern solution that was beautiful as well as practical, and often less expensive to manufacture. Although he was always identified with streamlining (the use of smoothed shapes that reduced air or water resistance), Loewy was never enthusiastic about that term, preferring instead to speak of simplicity. He praised light, airy, elegant, "thoroughbred" creations such as the bridges of Swiss engineer Robert Maillart. The worst design sin as he saw it was fussy, busy surface decorations, the principal shortcoming in the creations of the premodern era. If a form still looked disorganized or unfinished after simplification, he advocated encasing it in a shell for an appealing appearance. Many of the most recognized early Loewy designs, such as his streamlined locomotives, the refrigerators and ranges for Sears and for Frigidaire, and the tractors for International Harvester, left the underlying mechanical or engineering apparatus largely unchanged while sheathing them in a newly styled surface structure. Purists criticized such efforts as wasteful and dishonest, but Loewy felt that they were appropriate if needed to produce a pleasing final form.

There were other criteria in Loewy's aesthetics. Designs must avoid what he called parasitic aspects such as noise, bad odors, glaring lights, hard or rough surfaces, and harsh colors. Ease of maintenance and of cleaning, and built-in safety for use by careless consumers, also counted. The latest materials and current styles helped make a product modern and up-to-date, but consumers also had a legitimate psychological need for the familiar and the comfortable: the new must not be jarring or disorienting. The ultimate responsibility of an industrial designer was to create a product that would succeed in the marketplace, and therefore consumers must never be forced beyond the point where the novel and the known elements balanced. That point Loewy characterized by his famous acronym, MAYA—the Most Advanced Yet Acceptable principle. As a result, his work seldom veered to the extreme or avant-garde because the public would reject it and the designer would have failed his client. Loewy was renowned for his keen sense of how far he could push the public's taste, and he understood the need to preserve reassuring traditional elements in designs intended for the mass market. This combination of old and new appealed to American consumers—it made progress and novelty accessible but not discomfiting or threatening. It was because of his keen sense of just how far taste could be pressed, and where it was headed, that Loewy's firm gained such an extraordinary presence in the material culture of the twentieth century.

Loewy's autobiography is many things, among them a primary document of the American Century. He shows how American industrial design's mixture of economic, aesthetic, and psychological components fitted the era's prevailing political ideas. Through the Great Depression, World War II, and the postwar decades, the new culture of consumption lay at the heart of the American Dream and the hope to make that dream a reality at home and around the world. The belief that the fundamental purpose of life was to create an ever better material existence for more and more people became the animating force in the nation's conception of itself as a soaring consumer republic whose example the rest of the globe should follow. Plentiful and secure jobs, homes, automobiles, and more and improved goods accessible to all—here was a potent vision. Business promoted this doctrine to encourage growth, to overcome unions, and to win the Cold War. It had a powerful appeal to the American people, and in the decades after World War II it also proved alluring to other nations. Loewy's New York, London, and Paris offices led the way in extending American-style consultant industrial design to Europe and beyond. The French immigrant America had welcomed so generously became a proselytizer for the American way of life and a naturalized citizen of the United States in 1938.

Loewy was well aware of the political punch in the American Dream. Speaking at the Harvard Business School in 1950, he noted that "The whole world admires and envies American products, American appearance, American quality." The nation "should, and I believe will, take advantage of this receptive attitude," he forecast. He lamented the fact that "no one has yet been able to make [democracy's] high spiritual values of freedom, liberty and self-respect a 'packaged' item to be sold to the rest of the world." But consumer capitalism and American products offered "substitute solutions." "The citizens of Lower Slobovia may not give a hoot for freedom of speech," he asserted, "but how they fall for a gleaming Frigidaire, a streamlined bus or a coffee percolator." Here was the key to victory in the Cold War, and to the extension of a democracy of consumption at home.

Never Leave Well Enough Along concludes by placing industrial design in this broad economic and political context. After coyly saying that he, Walter (Walter Dorwin Teague), and Henry

(Henry Dreyfuss) helped a lady (America) across Style Street somewhat against her will, Loewy portrays their profession's role in the nation's consumer economy. Industrial design had the "social responsibility" of contributing to the "lowering of the cost of manufactured goods" and to economic growth. This would both "speed up employment and bring more essential products to the underprivileged classes." "This," Loewy concludes, "is democracy in action."

Although it is difficult to gauge the true economic contribution of industrial design, there is no doubt that Raymond Loewy and his profession played a significant role in furthering the consumer culture of the twentieth century. First Americans and then many of the world's peoples embraced the goals of endless growth and constant novelty in the material civilization of the modern era. As much as anyone, Raymond Loewy symbolized the vast social changes implicit in the triumph of consumer capitalism. This autobiography gives a rare, fascinating, inside look at one of the major sources of the cultural transformation that produced the global drift toward the American Dream. We could hardly ask for a livelier, more engaging, and better informed guide to how so much of the modern world crossed Style Street.

Glenn Porter

HAGLEY MUSEUM AND LIBRARY

NEVER LEAVE WELL ENOUGH ALONE

PART ONE

Chapter 1

CORPORAL LOEWY

July, 1914. France is at peace and PFC Raymond Loewy is in the army. He has just been commissioned a corporal in the 8th Regiment, Corps of Engineers. He is stationed at Rueil, a ten-minute tram ride from his home in Paris. He is all dressed up in his new uniform waiting for his overnight leave of absence. It is four o'clock and the gates will open. At five to four, the bugle calls "assembly." Why assembly at leave time? All rush to the center field of barracks grounds. Attention! A sickening silence and then the terse announcement: Mobilization has been decreed. All leaves are canceled. Orders to prepare combat equipment and to stand by. Then, August 2nd. *WAR!*

I had a very full war. I became a captain, on the General Staff of the Vth Army. I also spent several months in the hospital following burns by poison gas. I worked closely with the American troops in Champagne, built a luxurious dugout during the trench war period and collected seven medals including four citations for action in combat.

This dugout of mine was quite the thing. I was at the time sergeant in charge of communications for a segment of the front lines. My pal, Corporal Brunet, and I soon found in wrecked houses a couple of chairs, a piece of red carpet, and a slightly cracked mirror. On my first leave to Paris, I brought back an odd load: instead of the usual bottle of applejack, sausages, and chocolate, I lugged around flowered wallpaper, drape material, and tufted pillows.

The dugout was most elegant. On the lee side of the shell-bursts, we planted a flock of geraniums, and over the entrance was a sign which I painted in Spencerian calligraphy:

The place soon became popular with the brass. Whenever the front was calm, they would drop in for a chat and have a look at the latest copy of *Fémina*, or *l'Illustration*, and sometimes *Vanity Fair*.

It wasn't all velvet, however, and it is there that I won my first Croix de Guerre citation for "spontaneously leaving shelter under violent artillery barrages to repair communication lines incessantly cut off by explosions."

I made my own breeches, as the government issue was badly cut and I enjoyed going into action as a well-dressed combatant.

By the end of the war, both my mother and father had died and our home was completely disrupted. All three Loewy brothers received the Legion of Honor, for conduct in battle. My parents having left us no trace of assets, we found ourselves in need of making a living—immediately. My brother Maximilian had left for America, having been demobilized first. There he worked for the French War Commission and managed to support his family. Georges, a surgeon, became Chief of Clinic in a Paris hospital after having spent a year at the Rockefeller Institute for Medical Research in New York. In Paris, I tried to find some work in all sorts of engineering offices—to no avail. It did not matter much, as I was not anxious about staying anyhow, and I managed to get my passage ticket to America from the French government. As soon as I was discharged, I embarked on the S.S. *France* hoping to get a job with the General Electric Company in Schenectady. That opportunity had been arranged by my brother, the doctor, while he was in New York.

During the crossing, an unexpected thing happened that changed my whole future. An auction sale of small articles donated by the passengers was organized by the Captain for the benefit of the families of shipwrecked sailors. Having nothing to give, I made something—a pen-and-ink sketch, on the ship's stationery, of a girl passenger in sports clothes, attractive, walking briskly on the promenade deck. It was auctioned off that same night after the Captain's dinner, sandwiched between a copy of *The Nigger of the Narcissus* and a used alarm clock. Bids were sort of anemic, ranging between twenty and fifty francs. To my surprise, my sketch stirred up quite a competitive bidding and was finally auctioned off at 150 francs.

"Who is the buyer?" I asked later in the evening.

"Sir Henry Armstrong, the British Consul in New York. By the way, he says he'd like to meet you."

Next morning I was presented to him, a kindly, gray-haired gentleman of retiring nature whom I liked immediately.

"What are you going to do in the States, young man?"

"Work. Probably in a research lab at G.E."

"Do you like to work in a plant?"

"Not especially. But I have to make a living quick."

"Have you ever thought of doing commercial art work?"

"No; in fact, I don't think I can draw."

"I think you can. Why don't you try?"

"Perhaps I might."

"If you wish, I can give you a note of introduction to a magazine publisher, a friend of mine in New York."

"I'd be delighted, thanks ever so much."

The shipboard conversation with Sir Henry, who spoke fluent French, left me, if I may use the expression, at sea. The more I thought about it, the more intriguing it became.

I landed in New York in the fall of 1919 in my uniform of captain of the French army. It was rather worn, but well cut and neatly pressed. I stepped off the beloved S.S. *France* carrying one suitcase and a trenchcoat, my whole fortune—besides forty dollars in American banknotes. My elder brother Maximilian—then living in the States—was waiting for me. The day was a lovely September afternoon and we were happy to be reunited after four years of combat duty and numerous battle scars. Customs formalities were soon over and I was anxious to leave the steel pier of the Compagnie Générale Transatlantique to feel the soil of America underfoot.

Max and I took a cab, an enormous Pierce-Arrow limousine, that looked like the staff car of Marshal Foch, and off we were. Our short ride through lower Manhattan took us to "One-Twenty" Broadway. We rushed to an express elevator and soon landed, baggage and all, on the top story observation platform of the Equitable Building, forty stories over the new world. The view was breath-taking—I was speechless. The wonderful dreams that had kept me going throughout the hostilities had finally come true. This was life! All the sadness of war, its ugliness, its sordidness were fading away. I remained silent, as deeply moved as when I had watched, recently, in Paris, the parade of the victorious armies emerging in a blaze of battle flags from the Arch of Triumph. I could hardly keep tears from my eyes. My brother understood and left me alone. New York was throbbing at our feet in the crisp autumn light. I was fascinated by the murmur of the great city, accented by the frequent whistles of the Hudson River ferries and the frenetic staccato of the elevators' automatic machinery. After thirty years I still remember this moment vividly.

Awhile later, my brother came to me, silently held my arm, and said, "Raymond, this is your new home. Somewhere among these myriads of buildings, someone needs you, right now. It is for you to find him. Do not despair or lose your confidence. Americans are nice people. They will give you a fair deal. This is your new battlefield. This is where you will have to struggle for survival." I was entering America, and felt very humble, and very grateful. Somehow, I already sensed within me an undercurrent of faith and confidence. It seemed as if I could, in spite of the staggering emotional shock, faintly feel the pulse of America. I knew I would love my new country and that I would quickly become more than a guest.

I never expected that one could fall so completely in love with a nation and a people. I already knew that very soon I would thrill at everything American, love everything American, perhaps blindly, but fervently and forever. I still remember how thrilled I was at the sight of the flag. After all these years, every time I see the Stars and Stripes in the breeze, I get the same emotional feeling. Aesthetically speaking, it is perfect. One of the outstanding "designs" of all times. It may be that a nation gets the flag it deserves.

Maximilian and I returned to earth quite moved but happy. I appreciate what he did for me on that day. A sensitive man, he had, in a simple, beautiful manner, made me understand the privilege that was mine: America was to be my home.

We walked a block or so on Broadway, Maximilian hailed a cab, and we were on our way in the strange city.

"Where are we going?" I asked when we had been riding for ten minutes or so.

"Home, to the apartment."

"You mean, 30 East Sixty-eighth Street?"

"Yes."

"Then, we should have turned right at Thirty-fourth Street, shouldn't we?"

Maximilian looked at me a little puzzled and said nothing. A few minutes later, I said, "We should really leave Broadway and take Fifth Avenue, otherwise we are going all the way to Columbus Circle for no reason at all."

My brother began to realize that for a fellow who had never

been in America I seemed to know my way around pretty well but he remained silent. We reached Central Park and I said, "You see, now we have the park between us and the apartment. You should really tell the driver to take the next cross-park underpass at Sixty-fifth Street."

"How do you know all that?"

"I studied the map of New York on the ship during the crossing. I had little else to do for six days."

My brother knew then that young Raymond believed in preparedness. Also that the cab driver had understood West instead of East.

Before we proceed with the case history of Captain Raymond Loewy *versus* pauperism in the United States, let us consider the American scene circa 1919, as seen by a young Frenchman just released from four years of active warfare. After the first few days spent in utter bewilderment at the size of everything, the turbulence, the rushing pace, I began to settle down and try to understand what was going on. There was plenty to discover.

Owing to my inability to understand the inhabitants, or to make myself understood, the first impact upon me was naturally more physical than intellectual. Having no verbal opportunity to satisfy my curiosity through conversation, my urge to comprehend and to assimilate was diverted to the three-dimensional world around me. Furthermore, this attention was concentrated, at first, on the objects and implements in my immediate surroundings: a streetcar, lamp post, a subway ticket booth (turnstiles did not exist), coin telephones, soda fountains, cafeterias, etc. Awhile later I began to observe the structures themselves, whether skyscraper, hotel, or railroad terminal. Eventually, I graduated to the skyline and the wider horizon. But I would revert to immediate surroundings and soon I looked around again. The first impact was brutal. The giant scale of all things. Their ruggedness, their bulk, were frightening. Lights were blinding in their crudity, subways were thundering masses of sinister force, streetcars were monstrous and clattering hunks of rushing cast iron; it was terribly impressive, gigantic, restless, and supervolted. At close range, it was inharmonious and out of scale. At a certain distance, however, it seemed to be less disrupted. At long range, it definitely acquired a feeling of harmony; it began to make some sense. Whenever I returned to close-range observation I became mildly uncomfortable and I tried, subconsciously, to escape it. Now, after thirty years, I begin to understand my frequent "cruises" on a ferryboat to Staten Island, or up a skyscraper. These excursions had a soothing effect. As soon as it was viewed from a distance, everything began to get in scale and to blend together and I regained a feeling of calm. New York's skyline was not only magnificent, it was right. Often I would take a trip along Riverside Drive on top of a bus and discover that it was relaxing and satisfying.

Later I began to realize that what caused my uneasiness might very well be the actual form, sound, and color of things around me. I began to notice them more vividly, and the results were consistently a shock. There seemed to be a general trend to massiveness and coarseness that was unnecessary. There was no disciplined economy of means and materials. All this did not seem to check with the mental image of America that I had formed in my mind while in France. I anticipated things American as being simple, slender, silent, and fast. Well, I was wrong. They were fast all right, but they were also bulky, noisy, and complicated. It was a disappointment.

All this, naturally, was very much in the background as I was under the immediate pressure of having to make a living and I had no time to worry about all that. But it was there just the same, and its mere existence left an impression upon my mind and turned out to be a factor that, years later, influenced my whole life.

Why were products and things so ungainly? To understand this, one must realize that the industrial revolution of the mid-nineteenth century had displaced the craftsman, and that engineers were really mechanics. The first mechanical products were put together by men of ingenuity and resourcefulness. The prime objective was to make the contraption—whether a coffee grinder, a lifting crane, or a steam engine—work. Will it work? was the question. No one gave a thought to cost and far less to appearance. Products were "engineered as you go" and they betrayed this technique by their haphazard, disorderly look. By the end of the century, machines and products were more numerous, more complex, and engineers had been trained. Trained in everything but aesthetics. So when mass production appeared on the scene, the country became flooded with products, usually of good quality but clumsily put together and wasteful of labor and materials.

But somehow, some well-meaning "artistic" men had resented the disharmonious mess and decided to do something about it. Having no doubt admired the angels of Tintoretto, the garlands of Rosa Bonheur, or the stone lacework of medieval Chartres, they went to work and applied "Art" to the machine. These misguided decalcomaniacs went on an artistic binge that will be long remembered. With total disregard for the fundamentals of the problem, i.e., the improvement of the product itself, they started on their task of embellishment. The results were, if nothing else, quaint. So we had locomotives festooned with garlands of roses, steam rollers with pink angels, and coal stoves peppered with quails, butterflies, and nosegays of forget-me-nots. It marked the age of decalcomania.

CHAPTER II

As we go through this book, we shall illustrate, in the title pages of some chapters, various principles which affect the industrial design technique, expressed in typographical form. The title of this chapter, for instance, is a demonstration of the use of a strong diagonal design element. Such a device is widely used in all sorts of product designs. On labels and packages, trademarks or brand names can be surprinted over a background of text in contrasting colors. In department stores as well as other retail outlets, signs to identify the various departments can very often be displayed in an informal manner.

Chapter 2

ADOLESCENCE

I have been exposed many times during my career to the inquisitive minds of a group of brilliant specialists of the periodical press called "researchers." These restless men gather data for writers commissioned to do a "piece" for a magazine. Having been classified as suitable material for several such articles, it has been my privilege to observe the gentlemen-researchers in action. I am beginning to do some research myself as to what makes them what they are. Curiously enough, researchers seem utterly uninterested in such trite contributing factors to success as talent, or just plain hard work. This is too dull. They must know all about one's background, early environment, far, far down to adolescence and babyhood. In some cases, they went so far into my past that it got me quite interested in it myself. I began to be curious about the Raymond Loewy of those early days. I tried hard to remember more about some dim incidents vaguely sensed rather than recalled. It was somewhat like going through a homemade self-psychoanalytical process and I rather enjoyed it. Let me say that it hasn't clarified why I became an industrial designer. In fact, it rather tends to show that one can become a successful practitioner in spite of practically everything. For my youth looks to me like a heteroclite assemblage of crazy situations. Could it be that such a "climate" is a requisite to success in the profession? My researcher friends will no doubt decide the issue for me. At any rate, I recommend such an expedition *à la recherche du temps perdu* to anyone who hasn't a thing in the world to do. It is more than I can say about psychoanalysis—the real thing—which looks rather dreary. Every time I see a photograph of Freud I wonder how a man who spent his whole life tête-à-tête with sex can look that gloomy. Maybe he and I don't look at the subject from the same angle. But let us return to my retro-introspective trip through the foggy morass of Raymond's early days.

I was born in Paris in 1893. My father was a rather successful writer on the subjects of finance and economics. He was associated with several publications specializing in those fields. My mother, a solid, handsome woman from southern France, was chiefly busy trying to keep my two older brothers from getting involved with the young models of the Rue de la Paix. This was a full-time job, especially in the case of handsome Maximilian.

Mother, one afternoon, had seen two horse-drawn cabs collide on the Place de l'Opéra. One taxi staggered hesitatingly and turned over on its side. There was no damage but the passengers had to be extricated through the horizontalized door. The first passenger to come out was a lovely and rather undressed young lady laughing like mad and anything but frightened. The second rider to emerge was brother Max, looking very much annoyed. My mother, as upset as the taxicab, was furious, for Max at this very moment was supposed to be visiting an old aunt in Neuilly.

This aunt, incidentally, reminds me of my first experience at school. Old Aunt Bertha (she is ninety-seven) used to take me every day to a school not far from home, a kind of kindergarten for kids about six years old. My progress in learning how to read was disastrously bad and it puzzled my aunt no end, as I was considered as being quite normal in other respects. Aunt Bertha decided to investigate the matter with my teacher, Mlle. Hortense. Mlle. Hortense was a brunette, rather young and not unattractive. Unfortunately, she had quite a bit of dark fuzz under her nose.

"Tell me, Mlle. Hortense, what seems to be the matter with Raymond. Is he attentive?"

"Yes, he is very well behaved."

"Does he listen to you?"

"Yes, he concentrates well, perhaps more so than any other child in the class."

"Are you sure?"

"Absolutely; he keeps his eyes on me to the point of occasional embarrassment."

"Then what's the matter with him?"

"I just don't know."

It didn't take long for wise old Aunt Bertha to find out what seemed to fascinate me so, and interfere with my concentration; it was Mlle. Hortense's mustache. Back at home, she gave me a real talking to and went so far as to threaten not to buy me the promised Easter chick if I didn't stop looking at Mlle. Hortense's mustache. This, I felt, was undue severity. I decided I wouldn't take it and made a Machiavellian plan. We were then living on the ground floor of an apartment building, Avenue de Neuilly. I opened the window and waited for the cop on the beat to stroll by. Then I yelled like mad at the top of my six-year-old voice, "Help! Help! Arrest my aunt! Help! Arrest my aunt!" The cop rushed to the apartment door and pounded upon it with a night stick, coming bravely to the rescue of a threatened child. My poor old aunt was most embarrassed and she never tried again to interfere with my fun at school. I even got the chick and raised him inside the fireplace until he became a fat, lewd, and especially stinking rooster. One day even *I* had to admit that the beast had better be given away to the iceman. Which I did.

Dear Aunt Bertha is quite a character. I saw her this summer in the tiny French village where she reigns as a queen. She keeps the inhabitants in stitches with her wisecracks. To someone who asked her the secret of her vitality, she answered, "When you wake up in the middle of the night, just don't eat hot caviar." Which reminds me of Satchel Paige's answer to a delegation of strait-laced old bags who asked him about the same question. He answered, "I eat nothing but fried food and I never do anything in moderation."

Father, whom I consider a reasonable prototype for a gentleman, was born in Vienna, son of a family of educators, physicians, and other members of the professions. He came to France when he was twenty. Two important things happened to him in quick succession: he married my mother and became a vegetarian. The last was a major development of semicatastrophic proportions. Spending one's youth in company of a vegetarian father is no cinch. His brand of vegetarianism was of a rather peculiar character. At the table he would absorb nothing but raw fruits and vegetables, which can be quite a strain on the other members of the household. But I could not help notice that once in a while Father would return home early carrying a large package wrapped in black paper, red string-tied, and bearing a sticker marked OLIDA. He then would disappear into his study and lock himself up with the mysterious package, alone. It was puzzling to me, as Olida was the foremost *charcuterie*, or sausage shop, in Paris. It was only years later that I discovered the disconcerting truth: Father in secrecy was simply gorging himself with a couple of pounds of liverwurst, possibly in nostalgic remembrance of his Vienna days. However, this savage sausage-shock treatment nearly killed him every time—as it would probably kill a healthy young jaguar. Father kept up his periodical orgiastic sausage binge to the last. His alimentary facetiousness was evidently contagious, as the other members of the family had unusual dietary traits too.

At that time I used to go to Chaptal College, a respectable school a good half hour's ride by steam tram from my home. Each morning I was given two nickels for the round-trip fare. But I preferred walking back instead of taking the tram, regardless of weather, as I wanted to save a nickel to satisfy a sizzling desire. I was only ten years old but I already knew the meaning of passion. My passion was Biberin. No one but a Biberin addict could understand its ineluctable grasp. Biberin was a sweetened granulated powder sold in a little round box. The consumer was supposed to stir it in a quart of water in order to make an effervescent soft drink of indelible synthetic flavor. It was colored in the most appalling shade of green (for anise) or Schiaparelli red (raspberry). But we kids had far too much ardent fantasy to follow this bourgeois recipe. Our system was more spectacular. We would quickly gulp a pinch of the powder, and *then* drink the water. The resultant blast is hard to describe and harder to forget. The gases from the explosion would tickle our noses and bring tears to our eyes. It was wonderful. It was chemical ecstasy. I realize in retrospect that any child submitted to that kind of diet would either perish or get toughened enough to make a success in any profession.

We had another delight, a penny candy called the "Roodoodo," which I associate in memory with my Uncle Charles, a brave old soul. He had fought the Commune in the heart of Paris in 1871, and could not be restrained from narrating one particular battle in detail, over and over again, complete with vocal sound effects (cannon, bugle, etc.). This ordeal was pretty hard to take and I was torn between my duty as a good nephew and the unbearable boresomeness of the tale. One day, Uncle Charles, in spite of his earnestness, could not help but notice a certain coolness in his audience-of-one. He was anything but bright but I decided in despair to try an idea: I managed to convey to him the impression, subtly at first and with progressive clarity, that the gift of a box of Roodoodo might be good public relations. The sweet old warrior with a certain sadness in his heart finally caught on to the idea of a shakedown at the hands of his own nephew. The idea was simple, workable, and could be boiled down to a plain formula that even he could understand: I would listen attentively to the familiar tale on a time basis and the unit of charges was the box of Roodoodo, a sticky mass of synthetic jelly that kids sucked ravenously. It worked fine for many years and I learned all about the early days of the Paris Commune with a minimum of suffering.

Every time I think of communism now, I think of Roodoodo, which isn't a bad idea.

Another source of income was my publishing business. At the age of ten I owned and operated a weekly news magazine named *Le Journal de Plombières* made entirely by hand. Profusely illustrated, it had an impressive masthead with the statement "All foreign rights reserved." The price was twenty-five centimes (a nickel, at the time), the format small (2″ x 3″), and circulation at its peak, generally at Christmas time, reached ten.* This successful publishing venture made possible a veritable Roodoodo orgy every Saturday. Also, more matinees at the circus, for clowns occupied a great place in my life. I adored them, especially the wonderfully gay, witty ones like Footit and Chocolat, the subtle ones like Grock or Little Tich.

My father liked to play the piano. I suppose plenty of normal children have also been subjected to a well-meaning piano-playing dad. Usually, the musical repertory of one's parent would be limited to a small number of tunes that would be "executed" on occasions such as anniversaries, baptisms, weddings, and other turbulences. My father also knew a very few tunes, but unfortunately he played them not well and often. The only point in his favor is that he did not play loud, as his aesthetic sense could not bear it. I am a little ashamed to admit that I considered, at times, subjecting my own father to the Roodoodo shakedown so successful with Uncle Charles in order to insure him an audience.

* In April, 1950, I served as one of the three judges of the twentieth annual newspaper layout and typography award, known as the Ayer Cup. I recalled with a trace of wistfulness my early newspaperman days, so many years ago.

As far as I can recall, my love of speed was in me from the very start. Its first detectable evidence took the strange form of a Lentil Race. It also marks the young genius' first invention. The event took place in a classroom of the Collège Chaptal in Paris. I organized and ran the first Lentil Race in 1903 at the age of ten. It consisted in planting one lentil in a little vase the size of a bar "shot glass" filled with earth. The planted lentil was kept inside of our individual hinged-top desks, carefully watered daily except Saturday and Sunday. We started the nonstop race on a Monday morning and the finish took place ten days later at the noon recess. By that time, the lentil had sprouted and grown tall. At the finish, each stem was accurately measured among incredible scenes of excitement sometimes bordering on violence. I usually won the event as I had discovered a way to irrigate my entry over the weekend through an elaborate system of wicks and reservoirs while the competition was getting desiccated.

As I grew older, this love of speed took more active forms. One early experiment in that direction was made possible by the introduction to the Parisian population of the first "autobus." It created a sensation and I wanted to be part of it. So I would hide myself from the conductor and steal a ride between stops, crouching on the bumper, head jutting forward in the wind at the terrifying speed of twenty miles an hour. It was sensational. It was also very dangerous. It led to other forms of transportation autohypnosis, including my first bicycle, and eventually to my heavenly passion, the locomotive, of which we shall read later. But let's go back to the bicycle. For a long time I had saved money in anticipation of the glorious day when I could buy one and own it, forever. And now seems as good a time as any to introduce my second and last brother, Georges, distinguished surgeon, member of the American Academy of Medicine. One early autumn day Georges said:

"Why don't you save some money to buy a bicycle? You know, bicycles are fun in the spring."

"Do you think I could save enough to get one by April?"

"Sure. You know, I've got an idea."

"What's that?"

"You've noticed that when Dad comes back home, evenings, he reads the stock market reports in the paper, which probably bores him a lot. Why don't you offer to read it to him while he relaxes in his armchair? I guess you could charge ten cents a reading."

"You think so?"

"Why don't you try?"

So next evening I tried the idea. It was a success from the start. Under the emollient of my monotone reading Father would go to sleep in a jiffy. I could get him totally blacked out in slightly over two minutes. Hours later, upon awakening, he would pay the fee with a trace of embarrassment at his overrelaxed attitude, but everything went on splendidly. It was easy money and, under my system, both of us benefited. Father got very relaxed, and my bicycle dream became a possibility.

Among the rites preceding the Reading of the Market was one I especially enjoyed. I would bring Father his black patent leather slippers and select a pair of socks. I noticed that he was very absent-minded about the choice of socks. So after a while I started experimenting with some interesting color combinations. One sock black, for instance, the other dark blue. Noticing a total lack of interest for the subtleties of my tone blendings, I gradually became more bold in my experimentations. Mother ended it all when Father walked in the dining room, one evening, with a blue and yellow sock on one foot, a red one on the other.

By spring, I had saved enough to buy a bicycle; a ravishing red racer, resplendent with nickeled trim, brass bandings, and golden foliations. Then trouble started; Georges borrowed my bike and kept on borrowing it. He took it on Sundays and took it on holidays. Every time I wanted to have a ride, the bike was gone. I was furious and unhappy with a case of bike frustration. Then summer came, and with it our mother's birthday. Georges and I were planning a surprise—and it gave me an idea.

I was already a skilled cook, having watched the family cordon bleus for many years, in silent reverence. Often I had baked a cake myself, under the cook's tutelage, and I was getting really good. Georges was totally ungifted and that gave me a Machiavellian idea:

"Georges, we must have some surprise for Mother at dinner."

"Yes, we should, but what?"

"Let's buy some beautiful flowers, yellow roses, for instance," I said.

"How much do they cost?"

"About twenty francs, I guess."

"I have only four francs. What about you?"

"Three francs fifty. That'll never do."

"Any other idea?"

"Tell you what. Let's bake her a cake ourselves."

"I don't know how," Georges said.

"Never mind, I'll show you. Just do as I tell you."

I started my own cake, and when it was well under way I took Georges in hand. I directed him in such a fashion that when his cake was finished, it had about the consistency of a cinder block. The making of the cake was under the star of ill luck. Georges dropped it in a coal bin. It naturally didn't break and I thought it might make Georges suspicious. It didn't. By that time the "cake" was pretty dirty and I gave Georges a little help. We finally placed the block in a silver dish and decorated the top with a little bicycle made of squeezed-out sugar, which made Georges look thoughtful. Then we surrounded the cake with a deceptive pool of thick marshmallow. At the end of the dinner, at champagne time and before the toasts, the two cakes were brought in—in style. This was "our finest hour." Mother sliced my fluffy pastry first, I being the youngest. Georges's cake was next. It resisted my mother's efforts for a while and finally split up in several plasterlike chunks. There was a moment of great uneasiness. Father, as usual, tried to save the situation: "Georges," he said, "it was very nice of you to go to all this trouble to make such a nice, sturdy cake—and you did it all by yourself?"

"No," said Georges, anxious to share the credit with his younger brother, "Raymond helped me."

"Georges just tries to be nice," said I.

"Oh, you *know* you did help."

"Only at the end, when the cake was all but finished."

"What did you do at the end?" Mother asked.

"Well, the cake looked a little dirty and I helped Georges wash it with some soap and water."

Our finest hour was over. We all drank to Mother's health and left the dining room in a subdued mood. I am sure Georges was thinking of bicycles. So was I.

Like most young boys, I started a stamp collection. I liked stamps because they are often beautiful, colorful, and romantic. Zambesi and Ireland, Labrador and Ceylon, Costa Rica, Peru, Samoa and Greece. I could dream for hours, looking at my lovely stamps from far away. One series of stamps fascinated me: the triangular stamp of Cape of Good Hope. Soon I had the whole series. They were really splendid, assembled together in their varied coloring. I decided that no other stamps were as beautiful. Why then collect others? So I kept on accumulating the lovely triangles until I had an album full. The other boys were completely puzzled, but I didn't care. Mine was a strange stamp collection of several hundred stamps—all alike.

It was my mother's custom to spend the summer months with her three boys at Trouville, on the Normandy coast. These are unforgettable memories. The high spot of each day was the arrival shortly before noon of the New York *Herald* red Mercedes racing car bringing to Normandy the Paris morning edition of the paper at record speed. James Gordon Bennett had a very special interest in this service. It was his hobby. I was overwhelmed by the frenetic quality of the daily event. I knew I was witnessing the birth of something very important. That thundering Mercedes would appear in a cloud of blue smoke and dust among bursts of backfire. It would come to a gasping rest in front of the casino, the goggled driver covered with dust hamming it to the hilt and slowly descending from his perch to the admiration of the crowd. My great thrill was to feel with my hand the honeycomb texture of the radiator still hot and regurgitating; clogged with squashed butterflies, bees, and sometimes tiny little birds, holocausts to speed. There was about it all a fascinating fragrance of hot rubber, steam, castor oil, and gasoline. It was sheer voluptuousness.

During the winter, my mother used to take me with her to Nice, where we spent several months in a villa inevitably called Bella Vista. By the time I was fifteen, she thought I should take some dancing lessons and she made all arrangements at a respectable dance studio. This was the time when the "Missouri Waltz" was played all over France in a slow, sirupy beat. I took two or three lessons with a charming young teacher. Unfortunately, she had a passion, like most people from the South of France, for garlic-flavored dishes. Waltz and garlic didn't mix; it was very discouraging, and besides my heart was somewhere else. It belonged to a dozen or so magnificent creatures of such trim elegance and finesse that everyone and everything seemed dull in comparison. The creatures were resting nonchalantly in the roundhouse of the railroad depot. I was in love with twelve steam locomotives.

These lovely things were streamlined, as early as 1905. They were the magnificent engines *coupe-vent* of the PLM (Paris-Lyon Méditerranée) and they made railroad history. They are still referred to as among the most graceful locomotives ever built and they effectively influenced the rest of my life. I used to hang around the roundhouse and I soon became friendly with the crews, bringing them Corporal cigarettes and sometimes cigars. As a matter of fact, I spent so many hours at the roundhouse that my dancing lessons were increasingly neglected. To this day I have never caught up with dancing. But my love for speed and the "fast look" began soaring sky-high.

Then I became fascinated by a new form of sport just starting and called "flying." I used to go to the Paris park, the Bois de Boulogne, where I had seen, as early as 1904, the very first take-off of Santos-Dumont. The Brazilian inventor and sportsman would cover three or four hundred feet, barely off the ground, in his cellular airplane, the "Demoiselle."

I was entranced but unhappy. I resented the early airplanes because they were so ungainly. Most of them were downright ugly, looking as preposterous up in the air as a flying shanty.

But soon another "sportsman-engineer" who happened to be somewhat of an artist took off the ground the first graceful flying machine. His name was Hubert Latham; the airplane, a monoplane, the "Antoinette." This was a revelation. It looked light, racy, and as we say in French, "witty." Compared to the flat-footed crates I had seen lumbering in the air, it was dreamy. I became interested, started making my own designs in scale model form, and soon had one actually flying. Boys of today can't possibly appreciate what an achievement this was, forty-two years ago. Soon, a flying club was formed, and our friend James Gordon Bennett, of the red Mercedes, offered a grand prize, the Gordon Bennett Cup for model airplanes. My entry won handsomely and I became quite a personality among the teen-age set of the Bois de Boulogne. Soon kids asked to buy my plane in such numbers that I considered its manufacturing possibilities. I took design patents, registered a trade mark (Ayrel), and rented an empty stable on an exclusive street. Soon after, I hired two mechanics and a salesman, and we started production. It was all so very, very chic! The Ayrel Corporation, with headquarters at 235 Faubourg St. Honoré, was in business and its president was fifteen years old. The presentation of the product in a neat box was magnificent and the sales went well. Soon I learned all about business at the executive level. I had a knowledge of patent law, wholesale and retail sales technique, discounts and bookkeeping, public relations, advertising, and labor relations. The National Aeronautical League heard about young tycoon Raymond Loewy and organized a series of lectures in large cities to popularize the new sport of flying—with R. L. as part of the event.

This was at vacation time. We would go to a provincial town, pack 'em in the local theatre, and give a lecture about the new sport. Then I would fly my model airplane in front of the audience. The trick was to adjust it so that it would take off the stage, fly in a large U-shape circuit over the audience, and glide gracefully right back into my hand. The audience was thrilled and quite delighted. They had never seen anything heavier than air fly before. Then I would make a short speech and answer questions. This was good training for later years and helped me considerably to overcome a great shyness that had made my life miserable until then.

This lecture tour lasted a fortnight and I enjoyed it very much. Then back to Paris and to school. The Ayrel business thrived and demanded more of my time, in fact far too much, and my family decided that I should devote all my efforts to my studies. Being a sensible young man, I agreed. I sold my corporation to the salesman and went back to my studies. I had saved enough money to finance a two months' trip through Brittany during the following summer vacation. It also provided funds to build a model speedboat three feet long propelled by an electric motor and storage batteries. It had a well-designed hull and it was fast. With the

Ayrel II I won the Branger Cup for model boats on the lake of the Bois de Boulogne. It strikes me now that in constructing both the model airplane and the model boat I had given much more attention to appearance than the other kids. It was a great satisfaction and probably the subconscious beginnings of a process that was to become my profession.

My brothers were watching me closely, and they encouraged me to try new things on my own, to rely completely on myself, and to work hard. They were a good influence and I owe a great deal to their constructive tutoring. They had confidence in me and they gave me confidence in myself. They were appreciative of my efforts and I never felt that I was working in a void. Georges and Max established the proper climate for their young brother and they never let me down. I was living in an atmosphere of passionate research, fascinated by anything new, unusual, or merely promising—whether it was a new paradoxical theory, a different automobile horn, a witty expression of Parisian slang, the *mouvement Dada*, a new play by Rostand, or a décor by Diaghilev.

All the while, Father kept on talking about Theodore Roosevelt, why I do not exactly know, and I was beginning to fear that he did not know either. Somehow the greatness of T. R. never permeated through his discourses. Just the same, I was very America-conscious. The cinema screen had brought the American scene right in front of my eyes, complete with revolving doors, shootings, cross-eyed cops, meringue pie, and suspension bridges. I was not only interested, I was puzzled. America sounded so wonderful, so virile, so "modern." But at the same time it had utterly incomprehensible traits that left me puzzled for days on end. So it was a great joy when I heard that my first cousin, the

Abbé Labalme, was going to visit us in Paris before leaving for America. Here was my chance to find out more about the land.

Father Labalme, a Catholic priest in charge of a small parish in southern France, was a brilliant man and a gourmet. Six feet tall, handsome, with the physique of a Notre Dame football player and the wit of a boulevardier, he was my constant delight. Every visit from him was a treat. He had absolutely no ambition in the world but to stay with his flock (who adored him) and to eat and drink well. However, the bishop of his diocese had noted his brilliance and had selected him as a member of a mission destined to attend an ecclesiastic conclave in Canada. So here we were, wishing him a safe journey aboard the old S.S. *Gascogne* and a pleasant *pèlerinage*. The year was 1903. He promised to tell me all about his voyage, upon his return. When he left, my feelings were a mixture of admiration, envy, and impatience. Impatience to learn all about the Redskins, ice-cream sodas and skyscrapers, electric fans and switchboards.

You can imagine my joy when one evening, months later, as I was returning from school, I opened the door of the living room and there he was! He looked magnificent, tanned, healthy, and all smiles. I thought what a wonderful representative of France to send abroad! So human, so keen. This fellow helped make religion sound hopeful and accessible. I liked him. We all felt festive, restive, and ravenous. The dinner was Lucullan in every detail, and worthy of my cousin's reputation as an epicure. After supper, we asked all the questions in the world; it was enormous fun, and he had a great sense of humor. He was amused at the archaism of the French tongue as spoken by the French Canadians. However, what seemed to have impressed him more than anything else in America was the strange food consumed by the natives. Whatever the subject in discussion, he would eventually revert to the food habits of "the American" as something quite curious.

The impression I got was that besides known civilized foods, "the American" was fond of eating certain types of building materials, cardboard, etc. He described to us a small object, thin, about two inches by three inches, apparently made with pressed straw and sawdust. People ate it ravenously, he said, and its name sounded like Korn Krisp or something. Then he described another biscuit, very, very thin that looked and tasted like Kraft wrapping paper. Still another had the same color and texture as corrugated cardboard but was less pliable. He had visited an enormous factory in upper New York State where one of these food materials was mass-produced in front of his very eyes. In order to prove his point, he had brought back several crisp samples of this unbelievable product, in a tin box, and he made us taste it. Exhibit A looked like a tiny pillow. It seemed as if the little bundle were made of laundry string browned to a crisp. It was very strange, brittle, and tasteless, and we were convinced of the veracity of my cousin's statements. It was only years later when I lived in America that I remembered the incident and recognized the product that puzzled me so much: a wholesome, nourishing staple of American fare, still found in most homes. I also identified the various cardboardlike wafers still found nowadays in the bread trays of the Waldorf or the Plaza. In all fairness, I must say that he also described some luscious steaks and juicy pork chops that made us forget the cardboard foodstuff and the fried laundry string.

But after all these years in the States, I am still somewhat puzzled by the success of so many queer edibles, great favorites throughout the country. It often looks as if some manufacturer, having developed a good product of some kind, had hesitated when it came to marketing: Will it be breakfast food or cardboard containers? Hesitation about the fields, often totally divergent, in which he should, or should not, enter his products. I could make my point clear if I were allowed to mention some trade names. Take, for instance, some of our highly flavored—I was going to say perfumed—gelatinous desserts. Such a product made of the purest ingredients modern chemistry can evolve could have been just as successfully marketed in tubes and sold as men's hairdressing. Or else some kind of the bland, creamy pudding mixes, favorite of millions. As a cosmetic (sometimes called vanishing cream) and sold in pink jars, it would probably have made a killing. Vice versa, I know some dental creams that look and taste very much like the spreads they use on smörgåsbord appetizers at cocktail parties.

The life of a young man around 1905 was an exciting one. Can you imagine a young boy who in rapid succession sees the birth of the electric light bulb, the telephone, the automobile, the airplane, the cinema, and the radio? How could a child born in my time wish to become anything but an active participant in one of these new earth-shaking developments? I couldn't see it. I knew I wouldn't live happy in anything not directly connected with these momentous discoveries. From then on, I set my goal and I laid a very straight single track leading up to it. I have never left the track once; and in spite of what appearances may indicate, I like my single track very much and the scenery as I go along becomes increasingly fascinating.

But let's return for a while to my Biberin days. It was at the Collège Chaptal that I tried first to learn English. My teacher, the unforgettable Professor Hartman from Leipzig, had a strong German accent. He was lonely and irascible. He would explain words to us in staccato gestures that made everyone nervous, and stilted progress. Any amateur psychiatrist could have told us that the poor wretch was subconsciously trying to keep us from learning English. He succeeded brilliantly. Most sentences he would use as examples contained references to Bismarck or Kaiser Wilhelm. As French children we were rather on the prejudiced side and his choice would not tend to create what is called "a receptive mood." After six months I gave up English and took up boxing. I became quite good at it, and I acquired the reputation of being very fast. I also acquired a series of first-class shiners that were difficult to explain and slow to disappear.

At that time, I had two new interests in life: chemistry and horse races. Both ended in failure of the so-called abject kind. Take the former, for instance.

Georges, several years my senior, was a student in chemistry and together we had rigged up a small lab in a storage room off the entrance hall of our home. I was his assistant. Together we made some excellent flare powder and decided to prepare a large batch for Bastille Day, France's national holiday. We tried a small amount and it burned with a magnificent magenta flame. Unfortunately, a misguided spark fell in the large container. That flared up with a terrifying whoosh. Georges and I jumped out of the window (the apartment being on the ground floor). We were both intact, but the whole building was in a panic and the apartment was a mess. My chemistry days were over.

Mother, once in a while, would go to the races at Longchamps, and on rare occasions she would take me along. I was fascinated by the color, the pace, and the excitement. One certain vacation day I asked Mother for permission to go to Longchamps alone. She hesitated and finally gave in—with the understanding that it was an exceptional favor. She gave me the price of an entrance ticket and I left home in a state of great excitement, as I had a Plan. For a long time, my conception of heaven on earth had been a café in Paris then famous for its ice creams. It served nearly twenty kinds of ices, ice creams, sherbets, and spumonis. The name was Café Napolitain. My father used to take me there on occasion, but I never had my fill. I decided to take one long chance and try the races as a source of ice cream revenue. I spent two agonizing hours at the track, trying to pick the one horse on which I would bet the five francs I had saved. Finally, I spotted a horse whose jockey's uniform was described in the program as "cherry with alternate bands of lemon and coffee." This sounded just like a slice of spumoni. I bet everything on the horse, watched the race in sheer agony, and won my bet. I rushed to the teller's window, waited to be paid, and left for Paris by autobus, conditioning myself during the ride for the coming frozen orgy. I stayed at the Café Napolitain nearly two hours and gulped eighty francs of ices. I left for home shivering and nauseated. The return by Métro (the Paris subway) was an ordeal I haven't forgotten yet.

I never went back alone to the races.

In fact, nearly thirty-five years went by before fate and Florida conspired to get me trapped inside of Hialeah. There I was, in the company of my friends, looking absent-mindedly at the list of entries for the day. I like horses' names, so glamorous and imaginative. Diomed and Daedalus, Salstram and Sunstar, Ayrshire, Coronation, Ormonde, and Aimwell. Beautiful sounds for lovely thoroughbreds. Suddenly I discovered, sandwiched between Teddington and Galtee Mare, an entry with the unbelievable name of Louis Shapiro. Now that deserved serious attention: any horse that ends up, with all the names in all the world, by being called Louis Shapiro must have something unusual. We all agreed, and placed a rather large bet at very favorable odds. Louis Shapiro, as you correctly guessed, won the race.

C H A P T E R

Identification of a product or a corporation should be made instantaneous through the use and wide display of a potent trade-mark. This is best achieved with some abstract symbol, as simple as possible and of high memory retention value. The symbol above is cipher three in Bengali. It would make a good trade-mark.

SEX AND LOCOMOTIVES

I left Chaptal College when I reached fifteen to attend the École de Lanneau, a high-grade prep school for boys yearning to enter the École Centrale, France's famous technological institute. The École de Lanneau was located quite far from my home near the Porte Dauphine. The quickest way to reach school was fortunately the most pleasing to me: it was by train, real steam train. At that time Paris was encircled by an urban railroad belt called *La Ceinture,* a form of early subway whose tracks were laid in a deep open-air canyon. Stations were approximately two miles apart and, luckily, close to both home and school. The round trip to l'École became a daily joy as the motive power was steam. What a lovely situation! Twice a day I was in actual contact with a real train. Getting up in darkness in the predawn of a wintry night lost its sting. Filled with *café au lait* and brioches, I would walk to the Porte Dauphine station in the cold drizzle, rush down the slippery stairs, and wait on the platform for the morning thrill. How lovely it was to look far into the fog for the coming amber halo that would progressively brighten both my heart and the glistening platform. The onrushing train would decelerate fast in a shower of brakeshoe sparks, wheels grinding on sand, and come to a stop; gaily illuminated in the dawn, fragrant with smoke, steam, and burnished sand. I would climb on board a small compartment, smelling of damp whipcord and wet rug. As I would travel first-class, the compartment was usually half empty, and I would settle on the beige spotty seats, pull out my books, and go over my lessons for the day. Sometimes concentration was easy and sometimes not. In the latter case, I would read, absent-mindedly, the same equation time and time again. . . .

$$\sqrt{a - b} = x \quad \sqrt{a - b} = x$$

$$\sqrt{a - b} = x \quad \sqrt{a - b} = x \ldots$$

I would be vaguely conscious of some dim interference. . . . It was nothing unpleasant, just a sort of faint rhythmic beat. It wasn't the rhythm of the wheels on the rail joint. It was more "human." At times like these, the smell of the old coach—of sulphur and kerosene—seemed to acquire a different quality, more like a perfume. The rhythmic cadence decidedly was not in the wheels. Could it be inside of me? Being of an analytical nature, I was puzzled but observant.

$$\sqrt{a - b} = x \quad \sqrt{a - b} = x$$

$$\sqrt{a - b} = x \quad \sqrt{a - b} = x \ldots$$

I soon detected that the phenomenon invariably occurred when another passenger was riding in the same compartment. Also that it never took place when the passenger was male. On the other hand, it would become very acute if the passenger were both female and attractive.

This was strange. Strange, but not unpleasant.

Soon another disturbing factor appeared. While waiting on the station platform I would seem less interested in the thrill of the approaching headlight and more in the choice of the compartment I was to ride in. Until the eye-opening day when I hesitated so long in trying to find the "right" compartment that I nearly missed the train. "Right" implied the presence of an attractive girl. Very revealing. However, this last-second technique would sometimes land me right in the middle of a smoke-filled compartment full of rain-soaked gents and stinking to heaven.

But this female-passenger business had me puzzled.

Could it be that there were other thrills in life besides locomotives?

The matter was to solve itself one freezing winter evening as I absent-mindedly climbed into the nearest compartment for the ride home. Surprise! In it there were two passengers. A good-looking young mother in her early thirties and her six-year-old boy. Several parcels were in the baggage rack of the warm compartment (shoppers, obviously back from a trip to the Louvre department store). The windows were covered with frost, the lights golden, and the air perfumed. The kid was fast asleep stretched on the seat, his head resting over the lady's folded mink, a red balloon marked "Louvre" tied to his wrist and floating around gently. The mother was lovely with her cloche hat and its veil stretched over sensuous lips. Neatly dressed, blond, on the "plumpish" side, she gave the impression of a well-fed young quail. I often marvel at the photographic mind with which I have been blessed and that makes me remember so accurately details of my early youth. Well, the lady was busily knitting, Junior was asleep, and I was reading my textbook. The complete futility of my effort became obvious as I couldn't keep my eyes from the radiant creature. She noticed my restlessness with an amused glint in her blue eyes. I looked up: she was smiling at me. My face was no doubt very congested, as she paused in her knitting, smiled again, and said in a low, sweet voice:

"Raymond, what's the matter?"

I blushed to heaven and managed to ask, "Madame, how do you know my name?"

"You've written it on the cover of your book," she said.

I blushed further and took refuge behind my trigonometry. I was under a spell, trembling with unfamiliar emotions, and I kept on watching her from the corner of my eyes. She stopped smiling and looked mildly perplexed. Then she paused in her knitting and looked at her sleeping boy. She glanced back at me, hesitated a moment, and placed her knitting on the seat. She lifted her black veil just enough to clear the tip of her exquisite nose, opened her handbag, took out the mirror, and proceeded to wipe off her lipstick in slow, sensuous strokes. I could feel the warmth of her quick glances over the mirror's edge. Her eyes were liquid and very naughty. I could feel the ineluctability of some impending disaster and I shrank farther behind my book. It was of no avail.

Deliberately, she got up, straightened her blouse over her proud breast, stepped across the narrow aisle, and sat close to me. Wrapping an arm around my neck, she held my chin in her hand

and proceeded to blend a very, very slow, melting kiss into me that left both of us dissolved and breathless. An engineer might best describe it as a kiss of helicoidal characteristics with accelerating progression in depth. Away from technology, may I say that the flavor of Parma violets in April is also an accurate description. The train decelerated and came to a stop. We were in a station. I managed to glance out and recognized Porte Dauphine. I staggered out of the compartment, carrying overcoat and books under my arm, and stood on the platfrom, transfixed, dizzy, and shivering. She cleared a patch of frost off the window and looked at me. She was lovely and wistful. Behind her I could see the red balloon near the ceiling. The train started slowly, she blew me a last kiss, and disappeared into the night. I stayed there for a while, ravished and wondering. I never saw her again.

Slowly, I went home. I was puzzled. Could there be such unfathomable joys as this in life? I had better start and investigate. And what about my previous standards of bliss? What about the red Mercedes? The PLM engines? It all sounded so pale and far away. . . . I felt a sense of guilt, for I knew that I was going to sell my locomotives short.

That night the family noticed my peculiar behavior and started to worry—of all things—about my health! Mother thought that I was thin and suggested a good checkup by the family physician. In order to avoid the embarrassment of further discussions I thought it wise to simply agree. Yet I well knew that there was nothing wrong with my health. I had just been weighed and found wanting. Wanting more.

I stayed at l'École de Lanneau three more years. For the young reader aspiring to become an industrial designer who might be curious to know what my marks were, here they stand in their beautiful simplicity.

Higher Mathematics	—	Excellent
Trigonometry	—	Excellent
Descriptive Geometry	—	Excellent
Mechanical Design	—	Excellent
Chemistry	—	Zero
Physics	—	Zero
Philosophy	—	Zero
Languages	—	Zero
Literature	—	Zero

I believe some of my talent at calculus could be indirectly traced to my dear Uncle Charles. On my fifteenth birthday he gave me as a present an enormous old turnip of a watch. It could do everything—mark the time of the day, the day of the week, the stages of the moon—and it rang like an alarm clock. It did it not at all well. It gained two and one-half hours a day, skipped Wednesdays, rang the alarm forty minutes early, and displayed the moon in reverse. Finding the correct time was a theorem of Einsteinian scope that gave me plenty of opportunities to exercise my mathematical ability. I never worried much about the moon but a great deal about acoustics, as the watch was most noisy.

All in all, the machine was driving me to some form of mathematical insanity. One day it all solved itself as it fell accidentally, or something, into the Seine at the Pont Royal. Another thing that the dear old turnip wasn't doing well turned out to be swimming.

I had to announce the sad news to Uncle Charles, who became properly dejected.

My rather depressing marks in literature might partially be explained by a series of troublous encounters with my professor. Once this gentleman gave us the following test as subject for our essay:

> "Baron Eriago, in eleventh-century Scotland, incurs the displeasure of his feudal Lord. On order of his Lordship, Baron Eriago is thrown naked into a black dungeon filled with starving rats and abandoned to his fate." (Note: Describe the subsequent happenings as seen by an imaginary eyewitness.)

The following day, in class, each pupil read his piece. It was a mess and a mass of gory goo with nothing at all spared. Everyone had a grand time making with the dear Baron. When came my turn, I read my paper as follows:

> I have considered the theme of the essay and I have reached the following conclusions:
>
> a. No eyewitness could possibly give a visual account of what happened to the Baron, as the dungeon is described as being pitch-black.
> b. Could it be that through some disclarity in the language of the text what was meant was an "acoustical description"?
> c. If so, isn't the theme rather unfit as a subject to be described in cold-blooded detail by an adolescent child?

Having thus supplied my master with some timid suggestions about logic, language, and manners, I rested my case. The verdict was a resounding zero accompanied by a few remarks about the risks involved in being a *petit Blanbec*, the French equivalent for smart aleck.

Sometime later, our teacher gave us as a theme for an essay a discussion of the pros and cons of low-cost housing for a town of northern France. The dear professor insisted that what would count most was the freshness of the approach to the problem, in order to show our imaginative perception. This was a challenge. Having "freshness" in mind, I wrote this:

Dear friend:

While passing through Lille (Nord) I received your written request for an interview about the low-cost housing problem. As you know, I am on my way to Paris, where I am to be inaugurated on Tuesday as Prime Minister of the French Republic. The subject you ask me to discuss is fraught with such explosive political dangers that I would rather avoid making a statement at this particular time. I know you will understand my position and give me a rain check. I shall further request you to refrain from publicizing the contents of this note, which is of a confidential nature.

Please don't hesitate, whenever you come to the capital, to drop in at the Chancellery and say hello. I'll be glad to help you get that transfer you mentioned a while ago.

A cordial handshake and *à bientot.*

Yours sincerely,

R. L.

P.S. Sorry to learn that your wife left you. Hope your baby daughter will be over the measles soon. Cheerio, old boy.

As a fresh angle, that was a fresh angle. I got another zero for having avoided the subject. Avoidingly speaking, I consoled myself at the thought that what was good enough for the Prime Minister to avoid was good enough for me.

In the spring, my family decided to move to a luxurious new apartment building on Rue Georges Ville. Mother, exhausted by the stress and bother of moving the family, was very tired, yet she saw to it that everything was in order before leaving for a long rest in the country. Mother was not the one to take any chances by hiring too attractive female help for the household. Before she left her husband and three male sons alone for a month, she overdid herself in her selection. The cook was a ghastly dried-up spinster and the two maids were antiquated and blowzy. There was consternation in the male clan and brother Maximilian was especially *bouleverse* and fuming. But he was a man of action. A few days after Mother left, he had located some different material and the new staff moved in: three good-looking, healthy gals in their early twenties. The cook was especially husky: a solid Norman peasant with red cheeks and voluptuous hips. The two maids were cute, both blondes, trim and compact. One from Champagne, the other from Brittany. The Loewy household was contented. Home was cheerful with all these kids around, and, as far as I know, nothing off-color ever happened. We encouraged Mother to take a good long rest and installed ourselves in sybaritic bliss. Without being really aware of all that it implied, the presence of these attractive females seemed to make life more satisfactory. In my subconscious search for aesthetic perfection, the addition of these specimens seemed a step in the right direction.

Mother's fifth sense must have warned her that something suspicious was going on, for she returned unexpectedly. She was staggered when cute little Jeanne opened the door. When Julie came out of the linen room, Mother was increasingly disturbed. The final blow happened in the kitchen; Françoise was bent over the floor, scrubbing hard and unblushingly displaying the major part of a pair of very attractive thighs, with Max in attendance.

Within a few days we were back to normality. But for a long

time, Mother would watch the mail for the postcards that Jeanne, Julie, and Françoise would address to us from various parts of the country.

As I have mentioned before, my formative years were spent in vivid consciousness of America and things American. We received the New York *Herald* every day and my Sunday mornings were enchanted by Buster Brown and Little Nemo. I avidly read translations of Poe, Twain, or Whitman. But there was another vector among the forces that shaped my destiny. It was the *Personalitée Britannique*. I was very much smitten with some exterior expressions of what one might call the Britannism *Parisien*. Few young Frenchmen of my age were even remotely interested in it or, for that matter, in things American. I must give credit to my father and brothers for their intelligent initiative in bringing these new horizons to my attention. Through translations, I was just as familiar with Sherlock Holmes's latest resurrection as any English kid.

Having read Wilde's extraordinary statement that "other people are quite dreadful. The only possible society is oneself," I was spurred to find out more about the nation that could produce such interesting ideas. The dandyism of Brummell and the causticity of Byron completed the picture, however sketchy, that I had formed, in my adolescent mind, of the British. But it gave me a great lift. Here was a type of sophistication worth looking into.

I liked the elegance in depth of things and people British. No mere veneer, there. I liked their icy wit, their silver, their ponies. The fragrance of blond tobacco, the smell of varnish, a whiff of heather, the texture of their tweeds. Tweeds in subtle shades with intriguing names: Thames Green, London Smoke, Chelsea Bronze. And their magnificent gloves, their heavy crystal decanters. All quality in depth. It left its imprint on me and I realize it now. Among my pleasant memories of this period is that of a small English drugstore in Paris. Over its resplendent polished mahogany door sat an exquisitely carved and gilt coat of arms underlined by a sign in gold letters: APOTHECARY TO H.R.H. THE DUKE OF CONNAUGHT. It was perfect in its sober elegance. And those subtle men's perfumes from Jermyn Street: Penhaligon, Stephanotis. Exquisite blends of jasmine and lime, sandalwood and roses. I delighted at the Englishmen's cool indifference, glancing through their monocles at the exuberant parade of my Parisian friends. They were, as the French have labeled them since Balzac's days, *très Buckingham*. So terribly chic and aloof; mildly exasperating, but their performance was magnificent. I was also impressed by the passionate moderation of their dress and of their manners, the exquisite simplicity of details. It was sheer elegance, distilled elegance. All eccentricities were ruled out: too daring a lapel, too unexpected a color, too deep a contrast were considered marks of unbearable vulgarity. This frightening simplicity was relieved by subtle accents of texture: a black pearl, a large amethyst, a reverse cuff of grosgrain, a thin chain of green gold. They were quite extraordinary, really, phlegmatically staring at the world with a mildly astonished expression. Also their morals were different, according to incredible stories that went round the town. Attractive young ladies were known to have gone through an entire dinner in private dining rooms at Foyot or Larue without their escorts' making the slightest attempt at rape between caviar and pheasant. It was chic. Their attitude appealed to me, as I resented vulgarity in any form.

As far as the French influence was concerned, let us simply say that I was permeated throughout by French logic and its lightness of touch. I was charmed by Duvernois and Colette, Proust and Jacques Boulanger. In the graphic arts, my idols ranged from the preciosity of de Monvel to the virility of Cheret, the elegance of Lautrec. This was France's golden hour. So much wit, so much gaiety and finesse. Tristan Bernard, de Flers, Lavalière, de Croisset, Zamacoïs, and Diaghilev. . . .

My intuition brought me in contact with many a dilettante, regardless of his station in the social setup. My friends were likely to be a talented but unsuccessful writer, a trolley-car conductor, or the old fried-potato vendor at the corner of the block. All had intellectual integrity. Regardless of their economic misery they were happy people. They remind me of Shaw's remark: "I'll die denouncing poverty. Be a tramp or be a millionaire, it matters little which; what *does* matter is being a poor relation of the rich; and it is the very devil."

My conductor friend and the potato peddler did not consider themselves poor relations of the rich. They agreed with Shaw in the more sophisticated sense.

Chapter 4

FASHION ILLUSTRATOR

During the crossing from Europe on the S.S. *France* I had come across a very silly article in the real estate section of a New York newspaper, entitled "Focal Center of Gotham Society." It showed a plan of the East Fifties on which dots marked the sites of New York's socialite mansions: the Whitneys', the Astors', the Fricks', etc., etc. A careful analysis had established the fact that the "smartest" location in New York was near the corner of Madison Avenue and Sixty-eighth Street. This focal point was marked by a star. Surprise! The star was right next to my new home to be. Wasn't that something! Here I was, landing smack against the star. It did not take me long to figure out that I was only sixty feet west and forty-five feet south of Society. This is the closest I have ever been, topographically or otherwise.

My new home was a walk-up four-story apartment building, rather modern at the time. The moment I walked in, I started to get organized. Getting organized consisted in transferring the contents of my suitcase to the closet and getting acquainted with the kitchenette, as I planned to cook my meals and save a few nickels. I learned about the milkman, the laundryman, the A & P store. Then my brother took me on a tour of introduction to my immediate neighbors. As he was returning to France the next day, he wanted to make sure that I met at least a few people before he sailed. He well knew that my funds could not last much more than a week or so and that without any English, without a job, it was bound to be pretty rugged.

The first visit was to my immediate neighbor on the same floor, a distinguished and stuffy old lady who lived with her attractive teen-age niece. I still wore my Captain's uniform. One look at me was sufficient. I could tell that she wasn't going to invite me much. I was evidently catalogued as some kind of overseas wolf that meant trouble. The poor lady was so mistaken! I was timid with women to the point of real psychosis. (This cruel form of insane timidity made my life miserable for several years.) After a while we said good-by, and I made the error of respectfully kissing the old lady's hand. This gesture, so natural to a well-brought-up young Frenchman, was anything but a success. She probably classified it as being "slick" and during my two years' residence in the building she hardly ever spoke to me again. For a long time I resented her unfair suspicion. As far as her young virgin of a niece is concerned, she never even said hello. I wouldn't be surprised if she ended up as a tart in a lumberjacks' camp—upper Wisconsin. Such endings are often the sad lot of gals whose adolescent days are under the management of old bags such as my respectable neighbor of 3B.

Maximilian and I proceeded with our tour of introductions and our next stops were at the various tradespeople in the neighborhood. More particularly to the valet shop owned and operated in the same building by a Polish gentleman, Mr. Samuel Fodor. Mr. Fodor was a nice, fat, bald old gent, always smiling, with whom I was to have frequent and inextricably involved dealings as neither of us understood the other, especially when we happened to be speaking in the same tongue. In retrospect I cannot quite figure out why we should have got involved in such hopelessly tangled situations, as the problem was always the same. The problem was to have the suit I was carrying on my arm pressed, *not* the one I was wearing on my back. This we never quite straightened out. But Fodor was a kindly guy and an honest one. There were many sad days in which I had tried hard to get some design work without results. Back home, all tired out and depressed, I felt unbearably lonely. Sad winter days when life looked so hopeless to a fellow fed up with war, fun-starved and broke. When feeling like that I would put some dinner on the gas range and take my suit down to Fodor. After the cold blast of the street it was nice to enter the little store, all warm and cozy, full of steam and the smoke of Sam's nickel cigar.

He would greet me with a big, toothless smile, draw a chair near the steam pressing machine, and keep on working, silently. I kept on wondering whether or not the Topographical Society editor had taken Samuel Fodor into consideration when he situated his Society Center. If so, it must have required a half dozen Whitneys and a couple of Astors to re-establish the balance.

I would stretch out and relax, under the supercilious stare of the American Tailor's lithograph of the Men of Fashion. These magnificent specimens of manhood were invariably grouped in a bored cluster near a Georgian fireplace. They looked silent, unamused, and dignified in their high collars, and they all seemed intent on looking at me, from the black-framed picture on the wall. I couldn't help noticing that American Men of Fashion don't wear mustaches. The withering gents made me feel very mustache-conscious. Perhaps I should shave mine. Perhaps that's what scared the lady in 3B? Meanwhile, Fodor was still chewing his cigar and pressing away. Then he would finally stop and look at me, realizing that he should probably start a bit of small conversation. At this sure sign of impending disaster I would rush back upstairs and have a look at the boiled beef.

During the war, my brother Georges had been on the field staff of Dr. Alexis Carrel, the great French surgeon, Nobel prize winner, and codiscoverer of a new method of combat-wound treatment. When the American government invited Carrel to come to the United States in 1917 to establish a War Demonstration Hospital, he took along with him two of his young assistants. One was Georges, the other was Pierre Lecomte du Noüy. The WDH was operating under the sponsorship of the Rockefeller Institute for Medical Research. Many distinguished New York matrons joined the hospital's staff as benevolent nurses and operated in my brother's wards. When he returned to France at the end of the war, he left many friends, several of whom were desirous to meet Raymond, the kid brother with a good war record. Lecomte du Noüy chose to stay at the Rockefeller Institute to continue scientific research in physiology. I met him soon after I landed, and it left a deep mark on me. He was an exceptional man. With the physique of a movie star, the wit of Oscar Wilde, the finesse of Voltaire, du Noüy was the quintessence of subdued elegance. He was famous for his repartee and he could do practically anything he wished, gracefully and well. Pierre, then about thirty and of independent means, had had the most diverse experiences during his young life. He was the essence of versatility.

Pierre had been a successful actor in a *théâtre intime* in Paris. He had studied biology and physiology and he finally ended up as cowboy on a Western ranch. He became an expert at lassoing. I saw him one evening at his home jumping in and out of his whirling lariat with the skill of a rodeo winner and the grace of Nijinsky. Combined with the fact that he was in tails and never took off his monocle, it was startling to see the great scientist in action.

His romances were glamorous, recherché, and brief. They were also profuse, as Pierre had the charm of a Don Juan. Tall and lean, with the profile of a Roman emperor (the slender type) and the rhythm of Laurence Olivier, he was the quintessence of masculine elegance.

Nothing equivocal about Lecomte du Noüy. It was reassuring to see that such a "he man," as they said at the time, could manage to carry fastidiousness to such extremes without reaching the dangerous zone of effeminacy. At home, Lecomte du Noüy could see nothing off-color in arranging some camellias in a cup of crystal, smoking a rose-tipped Abdullah, or spraying sandalwood in his library. This was the man who rejected an easy life of social success and café society glamour for the solitude of the laboratory. He died in 1947, one year after publication of *Human Destiny*—one of the most important books of all time about God and the philosophy of religious belief. In Pierre I lost a friend who had influenced my life when I needed it most. He had helped me crystallize many fleeting thoughts, hopes, and desires for which I had had no criteria. He put everything in sharp focus. It became clear that one can lead a life of elegance and sophistication without running the risk of being tagged as affected, effeminate, or perverse. I never forgot the lesson taught by such a master.

After du Noüy, Samuel Fodor, my pants-pressing friend, was quite a transition. I liked him as well, in his directness and with his great wonderful heart. He had the effect of bringing me back to earth in a jiffy. Near him I would quickly realize that I had better make a living fast—or else. The transition from camellias, tails, and sandalwood to pants pressing was shattering but convincing.

So I rolled up my sleeves, took the regulation deep breath, and started to chase the almighty dollar. The first step was to look up some of my brother Georges's friends for guidance. One of them happened to be, in his dignified way, quite as extraordinary a person as Lecomte du Noüy. He was Herbert Straus, one of the Macy dynasty, and a gentleman. Speaking pure French, erudite and eclectic, he was blessed with a rare sense of aesthetics. We became friends quickly and thoroughly. Herbert Straus, like Lecomte du Noüy, was a man of infinite charm and elegance. He could wear, when dressed formally, a black silk-lined cape with the grace of Alfred de Musset.

At that time I was down to my last few dollars, rather bewildered by the fact that my inability to speak English made it very difficult for me to "get going." Herbert Straus—to whom I made it a point not to mention my distressed state of affairs—was subtle enough to read through my assurance that everything was just dandy. He went to great length to convey the idea that Macy's future was intimately bound to my consenting to give the store the benefit of my great talents. I might, for instance, contribute to the development of new ideas for the store windows. I had sufficient good taste not to pretend any further that I did not need a job, really, and so I accepted with joy and relief.

With things thus cleared up, it was agreed that I would start on the coming Sunday and that I would trim a test window during the day, so it would be ready by Monday morning. H. S. buzzed for someone and a few minutes later I was introduced to my new boss. At that time window-dressing people were not the sophisticated group they have become since. My new boss walked in, took a good look at me, and we both knew there was going to be trouble. He looked like a magnified version of a specimen I had seen preserved in an alcohol jar of some pathological lab. When H. S. made the irremediable *faux-pas* of telling the man that I would *not* be required to punch the store's clock, he looked at me as if he wished I were the clock and that he could punch the dial right out of my face. The critter commissioned me to do one key window on Herald Square. At that time the technique was to cram the window with a truckload of stuff, including half a dozen dummies on which tons of merchandise were piled up in layers. The result looked somewhat like a drawing room in the mansion of those two eccentric Collyer brothers of upper Fifth Avenue who never went out in fifteen years and were found dead under a piano or something. Besides being a bloody mess, it was poor merchandising. My feeling for simplicity blended with a dash of French logic indicated a different solution. That same night I worked out my scheme along simple lines. I dressed one dummy with a neat black evening gown, spread a luscious mink coat at its feet, and casually scattered around some accessories. Instead of the usual blaze of diffused floodlights, I left the window in semidarkness. The only illumination came from three powerful spotlights focused on the figure. The result was a contrast of violent highlights and deep shadows. It was dramatic, simple, and potent. It sang.

Around midnight, pleased with my experiment, I went home to get a few hours' rest before the store's opening in the morning. Then I returned to see the results. There was quite a crowd in front of *my display*. Mostly executives. They were talking in hushed tones as if the founder's daughter had been found raped in the window. It was a cataclysm, all right. Window dressers were already on the scene of the disaster doing rescue work by bringing armfuls of merchandise in order to repair the damage. My boss was in the window too, puffing and sweating in the floodlights. I quickly figured out that if I could get in the store fast, I might have a chance to reach the window, take him by surprise, and say "I'm through" before he had a chance to say "You're fired!" This is exactly what happened. To this day I have the satisfaction of being able to say that I've never been fired. Some may call it a close decision on a technicality, but it stands nevertheless. This was the first job I ever had; it lasted twenty-four hours, and I decided there and then that I would never accept a position with any firm again. I would be my own boss. If it hadn't been for the embryological specimen, I might still be slaving in abject resignation or be the subject of a Pulitzer prize play with a title like *Death of a Window Trimmer*. Well, thank you, you did me a favor.

One of my first visits was to a great American, Dr. Abraham Flexner, then Director of the Rockefeller Institute, later head of the world-famous Institute for Advance Study at Princeton, and Einstein's headquarters.

Dr. Flexner gave me plenty of very excellent advice of no immediate value for convertibility into ready cash.

Well, I had to do something quick and I remembered Sir Henry's letters of introduction. There were two: one for a Mr. Rodman Wanamaker, the other for a Mr. Condé Nast. I went to Wanamaker's first, which turned out to be a department store. I was received immediately and turned over to a young man by the name of Grover Whalen. Mr. Whalen wore striped pants, was very nice, and asked me to prepare a few drawings for an ad. They were well liked and I was promised more work. Everything was fine—until I realized that bills were paid by check at the end of the following month and that we were in the first week of October. This was getting serious indeed.

I went to see Mr. Nast. When I walked into the executive offices, I could sense that my uniform of a bemedaled French officer seemed to affect receptionists and secretaries favorably.

Mr. Nast saw me. He was most courteous and bored; but he spoke no French, and the conversation was a little strained. It was pouring outside and he kept on looking out the window and saying, "Nasty weather, nasty weather." Owing to my limited knowledge of English, I thought that rainy weather in New York was named after him, and it was very confusing. Finally, he placed me in the hands of a young, charming, and vivacious executive by the name of Edna Woolman Chase. She was encouraging and gave me some work, too. The drawings were not too bad, and—surprise—I was given a check in payment immediately. This was heaven! Also feminine intuition, I think.

I rushed out to the nearest bank to cash my check. Nothing

doing. They didn't know me and they wouldn't cash it. Two other banks, same results. I began to wonder by what mysterious process Americans managed to get hold of plain dollar bills, quarters, and nickels. Finally I was made to understand that what I needed was a bank account. I opened one, and the lucky recipient of my anemic favor was the Central Hanover Bank. My account has been with it ever since. During these thirty-one years these men have been more than bankers to me; they have been friends and advisers. On several occasions, such as the inception of our partnership, the period of our expansion, the establishment of foreign branches, their sound opinion and help were most valuable. I consider them a credit to the American banking system.

Wanamaker and Condé Nast were feeding me work steadily and I made new contacts: Pierce-Arrow, Butterick, etc. I worked very hard, but I was afloat. Through one of my clients I met Florenz Ziegfeld, for whom I designed some costumes. Then Sam Bernard and Irene Bordoni for more costume designs. But I did not want to be sucked in along those lines of art work. Magazines, especially fashion magazines, were more to my taste. Also department stores, such as Saks 34th Street, that kept me busy for a long time. Then around 1924 I met a wonderful fellow and a great magazine editor: Henry Sell. He was in charge of *Harper's Bazaar*. Soon I was in contact with the young lady fashion editor, Lucille Buchanan. Miss Buchanan (Tookie, to her friends) is one of the few persons I have met in my life gifted with an absolutely sure taste and keen style sense. Extremely witty and gay, Tookie was a pleasure to work with and a joy to look at. We used to have lunch or dinner together once in a while, and I loved it. My English improved fast as she helped me a great deal. From the day I landed until the *Harper's Bazaar* days, my private life had been one of utter solitude. Never a date, never any fun. Just hard work, day and night, season after season, year after year.

Lovely Tookie understood my lonesomeness and tried, once in a while, to present me to her friends. I remember the wonderful times we had in Chinatown, like the two crazy kids we were. We liked to watch Chinese plays, eating sunflower seeds by the bagful, completely enraptured and in a deep fog as to what all the acting and puffing was about. These three years of adorable friendship left a mark on me; I am still influenced by her sophistication.

By that time, America seemed less strange to me as I slowly began to be assimilated. Some of the things that shocked me at first became progressively less startling. I even got used to seeing people putting sugar in mayonnaise and mayonnaise over peaches.

My work was progressing beautifully and I worked like mad. I moved my studio apartment to 52 West Fifty-seventh and worked literally night and day. Three or four times a week I would stay at my drawing table until daylight. I will never forget the early noises of New York's dawn. The resounding sound of ash cans being emptied and dropped back on the sidewalks, the horses' hoofs on the hard pavement (it was the milkman), and the hurried steps of some humble toiler going to his job. Once in a while I would look across the street at a group of strange and very ugly people that puzzled me. Four guys playing cards all night long, in shirt sleeves, in an empty loft. A couple of fast-looking dames would supply them with highballs, and smoke sprawled over an old leather sofa. One day they all disappeared. I read in the papers that a guy named Arnold Rothstein had been taken for a ride. That turned out to be one of my card-playing friends across the street.

I liked to relax for a while and look down in the street at these early New Yorkers. A drunk staggering back home, a tired call girl back from a night, carrying a brown paper bag in her hand—probably a pint of milk, ham on rye, and a dill pickle. (Also what looked very much like a pair of black panties.) The cop on the beat would yawn, leaning against a building, bored and sleepy. I would generally work with the radio on—a good way to pick up some English in a subconscious way. Also some unexpected broadcast gems. Once I heard Mayor Jimmy Walker speaking late one evening after a banquet or something to the New Yorkers during the Seabury investigation of graft among New York courts. Jimmy said, "I am accused of having bad judges. Why, I've got the best judges that money can buy!"

And the tunes of the time: "Sweet Sue," "Harvest Moon," "Old Man River," "For Me and My Gal" . . .

Chapter 5

Unusual arrangement of component units in the design of a product is of value providing it is consistent with function and economy. This page shows the possibilities of such a design approach in typographical terms. On a label, a carton, the side of a delivery truck, its advantage is to attract attention in a manner not inconsistent with aesthetics.

THE CRUSADE

One day in 1927 I met a man whom I shall never forget: Horace Saks, then President of Saks and Company. It was Saks 34th Street, as there was no other Saks store in New York—or anywhere else—at the time. Horace Saks, a great merchant, was also a man of vision and kindness. I began making sketches for the store's advertising, and I discovered, to my delight, that he was receptive to advanced layouts and unusual typesetting. This was very refreshing. I became more daring, and he often congratulated me on my work.

One evening, as I was taking instructions about the next day's ad from the Advertising Director, Horace Saks asked for me. We drove uptown together and he announced the big news: Saks was to have another store uptown on Fifth Avenue at Fiftieth Street; it was to be the last word in store design, and it would be called Saks Fifth Avenue. We had a grand time talking about the future, the wonderful things that could be done, and he said he wanted me to think about it.

While the store was being erected we often had informal talks about the new setup. I told him that I felt we should take the opportunity to create a new kind of store climate, that we should establish unusually different standards in every phase of its operations. A new store philosophy would have to be developed. One of my first ideas was to have a high level of employees. All members of the sales force, whether male or female, should be carefully selected for neat physical appearance and courteous inclination. They would be dressed simply and well. Such details as wrapping paper, cardboard boxes, paper bags, etc., should be designed attractively. At the time I was mildly shocked by the appearance and manners of elevator operators in general. They were rather untidy, their hands not quite as neat as they should be, and not specially courteous. I felt that one gets quite intimate, whether one likes it or not, with elevator operators during a crowded ride at rush hours. The least a store owner could do, in deference to his customers, was to see to it that the operator—male or female—should be correct, polite, and neat. H. S. agreed and I suggested designing a special uniform for the crews. This uniform created quite a sensation when the store was opened. The men wore neat dark suits with white piqué Ascot ties and white gloves. Within a short time it was copied by other stores all over America.

I prepared the whole advertising campaign that preceded the inauguration and did most of the art work for several years, until Horace Saks suddenly died. This was a blow to me, as we had a real affection for each other and a great deal of mutual respect.

My good friend Adam Gimbel took the helm. He had the

same store philosophy as Horace Saks, and he developed it further still. He made of Saks Fifth Avenue an institution known all over the world.

Louis Gimbel, in those years, was Adam's assistant and a charming young man about my age. He spoke French fluently and I used to see him often at the store. We had many things in common, particularly an affinity for Cuba and the Cubans. About that time, I began to go to Havana every winter for a short vacation, mostly for rest, unlike the average tourist. I would go alone, and stay at some small colorful hotel, alone, for two or three weeks, and I loved it. I had such a fondness for the island that during the revolutions of the early thirties I went back just the same even though tourists were outnumbered by machine gun squads. These were scattered all over town, the men sitting on the sidewalks at street corners.

At night, the sound of revolvers and rifle shots, the roar of fire engines, blended curiously but well with some slow rumba played behind the steel-shuttered front of a corner bodega.

Shortly before the revolution, Louis Gimbel and I discovered Constantino Ribalaigua. Constante was the barman at la Florida, a café in Havana, and his Daiquiris were unsurpassed. Louis, one day, had a grand idea: he was going to wean him away from Cuba, bring him to the States, and the three of us were going to be in business. He had it all figured out: he was to be the angel, Constante would mix 'em, and I would decorate the place. It all sounded very good at the time, in the noon sun of the Caribbean plus a great many frozen Daiquiris (made with a dash of maraschino). The atmosphere was not especially conducive to coherent thinking, as la Florida is probably the noisiest place on earth. The little café, with its fragrance of ripe pineapples, limes, and cigars, is all marble and mirrors and the ceiling forty feet high insures shrill resonance to the assortment of sonorous blasts that nearly shakes it to splinters. As it has no outside walls, windows, or doors, all the noises from the crowded narrow street just come right through. So does part of the agitated overflow from the sidewalk, including yelling and shouting lottery ticket vendors, peanut merchants, etc.

Eight athletic bartenders shake in rapid staccato gallons of Daiquiris for the back-slapping, hard-drinking Habaneros while four Afro-Cubans shake a loud rumba right in your ear. Dozens of loosely bolted Wahwahs (the tiny Cuban busses) roar and screech in an orgy of stripped gears; rattling trolleys clatter and shriek, all gongs clanging; it is plain mad. It sounds as if the entire Cuban transportation system had been temporarily rerouted through la Florida. To make it more frenetic, three enormous tornado-like electric fans (the kind they probably use at Langley Field's wind tunnel) pivot slowly and project recurrent hurricanes that raise waves in your glass. Even calm Constante, at times, gets a bit glassy-eyed as he gets noise-dizzy. One can understand how Louis, a good salesman when cold sober, managed in this atmosphere to get Constante to sign on the dotted line. When Louis returned to New York, sun-tanned and calmed down, things took a different aspect. So Constante stayed in Cuba, where he belonged, with his family, who never wanted to come in the first place, and this was the end of Louis and Ray's New York Daiquiri Bar.

I saw Constante recently, after nearly twenty years, and together we had a coffee cocktail to the memory of those happy days. La Florida had not changed. Thank heaven nothing had been streamlined and Constante never heard of Le Corbusier. The black bean soup is still as delicious and indigestible and the menu is as colorful as ever.

For the sake of Cuban-American friendship, and for the preservation of a very likable type of Hispano-American literature, I copied the menu of la Florida. With Viola, my wife, I preserved its fragile orthography through Wahwah ride, plane flight, and customs inspection. Here it is, undisturbed and authentic:

LA FLORIDA

LANCH MENNU

SCAMBBLED EGGES

AMBURGŪESE WITH ONNION

PORKSOP WITH BADKED APLE

COLIFLOWER

ERTITSHOK

WATER SCREWS SALADE*

DESERT

DIPLOMATIC PUDIN

DRIKS

CUBAN COFE AMERICAN COFE

JIBOL

(The letter *J* in Spanish is pronounced somewhat like the silent *H* in English, partly explaining this curious spelling of the word "highball.")

* Watercress salad.

Louis, during the war, joined the Air Force, became a colonel, and helped organize the Air Transport Command. He asked me as a personal favor to design for him an emblem for the ATC, which I did. It became known the world over during and after the war.

Louis was killed during the first year of hostilities. Cubans have not forgotten him. Neither have I.

Paul Bonwit, whom I had met shortly before, was planning a move similar to Saks'. He was going to build a new store at Fifth Avenue and Fifty-sixth Street and he discussed the matter with me. I saw an opportunity to contribute some constructive thoughts, and I became their advertising free-lance artist. This also was a success, but I wasn't really happy. Financially speaking I was successful, what with Bonwit Teller, Shelton Looms, magazine illustrations, etc. But I felt frustrated. Yet I kept on plugging—and hoping. It seemed to me that all this concentrated effort would eventually bring results, that it would make it possible for me to do, someday, the things that I felt it to be my destiny to accomplish. So I kept it up, working twelve to eighteen hours a day and on occasion reaching the fringe of exhaustion. I had reached the saturation point and I felt I needed some exercise. So I took a few days of absolute rest and resumed boxing, which was my favorite sport while in college. I used to exercise two or three times a week with that good instructor and grand referee, Arthur Donovan. It kept me in shape, fit and fast on my feet; just what I needed for the job ahead. Once in a while I would go to the New York Athletic Club for some fencing lessons.

Meanwhile, Tookie was initiating me to the delights of American humor. Through her, I acquired a great fondness for its dryness and its concise qualities. Often she would telephone me from the *Bazaar* and tell me the latest joke. Some, she admitted readily, were not new, but they were good material for a Frenchman trying to get the flavor of American humor. This early conditioning made it possible for me to appreciate the humor of later masters such as Virgil Partch, Peter Arno, Thurber, and above all, the weird genius of Charles Addams. About 1928 it was my good fortune to meet a great cartoonist and humorist, a delightful man and a sensitive one. His name was Ralph Barton, and he knew well both Tookie and Henry Sell. Ralph's drawings stood out among others like islands of tone and line purity. So I felt that Barton was doing in the illustration field what I wished to do in the three-dimensional world. It had everything I liked: grace, fluidity, and charm. Aesthetically speaking, it was reduction to essentials, without ever reaching dryness and sterility.

A few years later, Ralph committed suicide; a loss to his friends and to the world. Just before the end he wrote an unforgettable farewell note which, I understand, has never been published. Despite the gruesome aspects of the subject, I believe it has such depth, such subtlety and heartbreaking humor that I think it ought to be printed here:

OBIT

Everyone who has known me and who hears of this will have a different hypothesis to offer to explain why I did it. Practically all of these hypotheses will be dramatic and completely wrong. Any sane doctor knows the reasons for a suicide are invariably psychopathic and the true suicide type manufactures his own difficulties.

I have had few real difficulties. I have had on the contrary an exceptionally glamorous life as life goes and have had more than my share of appreciation and affection. The most charming, intelligent, and important people I have known, and they liked me, and the list of my enemies is very flattering to me. I have always had excellent health, but from childhood I have suffered from melancholia, which in the past few years has begun to show definite symptoms of depressive insanity. It has prevented my getting anywhere like the full value out of my talent and for the past three years has made work a torture to do at all.

It has been impossible for me to enjoy the simple pleasures of life that seem to get other people through. I have run from wife to wife, from house to house, and from country to country, in a ridiculous effort to escape from myself. In doing so I am very much afraid that I have wrought a great deal of unhappiness to those who have loved me.

In particular, my remorse is bitter over my failure to appreciate my beautiful lost angel, Carlotta—the only woman I ever loved and who I respect and admire above all the rest of the human race. She is the one person who could have saved me, had I been savable.

She did her best.

No one ever had a more devoted or more understanding wife. I do hope she will understand what my malady was and forgive me a little.

No one is responsible for this—no one person except myself.

If the gossips insist on something more definite and thrilling as a reason, let them choose my pending appointment with my dentist or the fact that I happen to be painfully short of cash at the moment. No other single reason is more important or less temporary. After all, one has to choose a moment and the air is full of reasons at any given moment. I've done it because I'm fed up with inventing devices for getting through twenty-four hours every day and with bridging over a few months periodically with some purely artificial interest, such as a new "gal" who annoys me to the point where I forget my own troubles.

I present the remains with compliments to any medical school that

To this last self-drawing made the day before he committed suicide, Ralph Barton retained his sense of graphic subtlety. And evidently, as seen in the background, women were very much on his mind. *Courtesy Henry Sell.*

fancies them, or soap to be made of them. In them I have not the slightest interest, except that I want them to cause as little bother as possible. I kiss my dear children—and Carlotta.

Ralph was found in the morning by the housekeeper. He was seated at a small table, before a mirror. Opened on the table was an anatomical chart of the human body, a pencil mark around the heart. He had shot himself at that same spot, with great precision, and apparently with infinite calm.

Boxing and fencing were good diversion and I worked harder than ever at my illustration work—but still I was not happy. Somehow, I was more and more concerned about the amazing things that were sold all around the stores where I was called every day in business.

I was in a constant state of admiration for the mass of products resulting from superior American technology and drive. I just couldn't believe that there could be such a wealth of productive genius. The country was flooded with good, inexpensive things that practically anyone could afford to buy, products which, in Europe, would have been considered sheer luxury; production of such incredible magnitude that it overflowed normal sales outlets and sought new sales channels, such as cigar counters, newsstands, and—most notoriously—drugstores. For a Frenchman, accustomed to the severe professional dignity of the French apothecary shop, this was a shock. Instead of a dark stuffy little pharmacy reeking of carbolic acid and wintergreen, here was a flashy, dazzling store crammed full to the ceiling with everything in the world from aspirin to roller skates, a garish phono blaring away "Dardanella," and the smell of fresh coffee and Pinaud's lilac trying to drown out iodine and cheese sandwich. My French friends, those who have never visited the States, still can't believe that even in a small village one can send Junior to the drugstore for an ounce of mercurochrome, two records by Spike Jones, an automobile jack, and a pint of chop suey. And that's nothing. Wait until we industrial designers get busy on the drugstore problem. You may end up by going in there to get an electronic haircut, while looking at Monsignor Sheen delivering a video sermon nine hundred miles away. By that time, live-wire undertaking establishments may have added to their regular line, as business boosters, such sideline articles as fast motorcycles, glazed bathtubs, or folding ladders. Nothing like a quick turnover, you know.

To return to our subject, I was much impressed, as stated before, by the quantity of products mass-manufactured. I was equally affected by their superior quality. With a few exceptions, the products were good. However, I was disappointed and amazed at their poor physical appearance, their clumsiness, and—to be frank—their design vulgarity. Here were quality and ugliness combined. Why such an unholy alliance?

Ugliness of color, of mass, of detail. Once in a while, a product would be more cohesive in its design. But then it would be utterly spoiled by a lot of applied "art": a mess of stripes, moldings, and decalcomania curlicues that would hopelessly cheapen the product. It used to be called gingerbread. (Now we call it schmaltz, or spinach.) What's more, all this corn was expensive: it did not generate spontaneously; it had to be painted on, etched in, stamped out, slid over, pushed out, or raised up; baked in, sprayed, rolled in, or stenciled up. It meant unnecessary work and, therefore, parasitic cost increase to the consumer. I was shocked.

I felt very unhappy about the whole business. The uncomfortable feeling I had when I landed in America became more acute. What was the matter with manufacturers? Why couldn't they see the light? How long would they fool themselves and the public? Why manufacture ugliness by the mile and swamp the world with so much junk? Why couldn't what used to be called "the machine age" generate simple, straightforward products and contribute a little beauty to the world? Why couldn't society be industrialized without becoming ugly? This sort of thinking kept me awake at night. I became terribly restless. I felt I should try to do something about it. I became acutely dissatisfied with my work as an illustrator, and this was unbearable. I was making an excellent living for a relatively young man of thirty-three, probably thirty or forty thousand a year. But I didn't care to go on thinking up department store layouts for diaper ads when there were myriads of nice, decent products being kicked around and abused, shrieking for help and deliverance from the arty manufacturer. I seriously considered two possibilities: a. I would return to France, buy a farm, and forget about the world, in the company of a flock of Irish setters. b. I would start a one-man industrial crusade under the aegis of good taste, and do it right here in America.

This was to be a momentous decision and I decided to take my time, to weigh every factor, and then decide in a realistic, unemotional manner.

The first signs of market saturation were felt about 1926 by the buyers of some of the department stores for whom I worked. Automobiles were ugly to the point of being quite repellent. How long would the public put up with it? American tourists returning from abroad would bring back Rolls-Royces, Hispano-Suizas, Bentleys, and Voisins that looked low and graceful. People liked them. Imported German or Swiss machinery, French hardware or Italian dynamos had slender, simple lines, that were admired by many on this side of the Atlantic. So I began to realize that the trouble with American products might conceivably not lie with the consumer. It was more likely lack of vision on the part of the manufacturers. In other words, they underestimated the potential good taste of the American buyer. The general economy seemed to indicate the imminence of the expected competitive era, and I felt that manufacturers would have to make special efforts to sell their wares.

The situation appeared to filter down to some simple facts:

By 1919, the time of my arrival in New York, products were at the end of the decalcomania stage—not quite so clumsy as in the nineties but still very crude.

Then America entered a decade of gigantic industrial expansion. The whole country became electrified (literally speaking) and tens of thousands of miles of highways were built to accommodate a flood of motorcars. Business curves were soaring, and manufacturing too. It was a sellers' market and an era of unprecedented prosperity.

The entire production was absorbed, by an eager public, on performance alone and in spite of appearances. The country was flooded with refrigerators perched on spindly legs, and others topped by clumsy towering tanks. Typewriters were enormous and sinister-looking. Carpet sweepers when stored away took the greater part of a closet, and telephones looked (this is no pun) disconnected. I felt that the smart manufacturer who would build a well-designed product at a competitive price would have a clear advantage over the rest of the field when things would become tough.

So here we were: saturation was at the door; competition would become fierce, good design could help sales, manufacturers could be convinced—and I was the one to do the job, both the designing and the convincing. Besides, I had become attached to America; I felt I belonged here. I liked everything about it. I remembered how happy I had been every time I returned from Europe, the wonderful feeling when sailing through the Narrows on the deck of the *de Grasse*, the *Paris*, or the *Rochambeau;* the emotional shock that made me end abruptly many a conversation with a choke in my throat and the feeling that my eyes were uncontrollably becoming too wet for comfort. (In later years I took the precaution of wearing dark glasses to avoid the embarrassment I knew I could not avoid during that part of the voyage home.)

How could I leave all this forever? It seemed impossible. And I thought that if I ever succeeded in my industrial design venture it could be enormously exciting. Why not try for a year or so?

The crusade was on!

PART TWO

THE DUPLICATING ANGEL

Whether or not the Lord heard my prayer, an angel just then flew out of the sky and landed on the terrace of my small penthouse. Of all things, the angel's name was Sigmund Gestetner. He was a plump, very nearsighted angel and his harp happened to be a mimeograph. He was also a kind and progressive fellow. Sigmund, a successful British manufacturer of duplicators on a short visit to America, showed me a photograph of his product.

"Mr. Loewy," he said, "do you think you could improve the appearance of this machine?"

"Certainly."

"How would you go at it?"

"I would have to see one first."

"I will have one sent here to you, tonight. Then what?"

"I will start working on it at once."

"How much will it cost?"

I hesitated a second. "Two thousand dollars."

"And if I don't like it?"

"You *will* like it."

"I believe I will. Yet, supposing I can't use the design?"

"Then I will charge you for the cost only, say five hundred."

"Righto, but I am sailing in five days and I'd like to see the design before I leave."

"Okay, you'll see it within three days."

When Sigmund left I was frantic with joy. I opened a demi-bottle of Moët and drank to my angel, to the world, to me. I also remembered my first communion days and I thanked my wonderful Lord. It was heaven!

Then action. I ordered by telephone one hundred pounds of modeling clay, modeling tools, and a floodlight. I took a cab, rushed to a Forty-second Street war surplus store, and bought a large tarpaulin which I stretched over the pale beige carpet of my living room. I told my Filipino combination cook-valet-chauffeur-and-secretary to take all phone calls and not to disturb me till someone would bring a machine. Then I lay down and went to sleep.

The Gestetner duplicator was no better and far worse-looking than many other products then on the market. It was a well-constructed machine that would make mimeograph copies of whatever one might have the impulse to mimeograph (an urge I have not yet experienced). Unwrapped and standing naked in front of me, it looked like a very shy, unhappy machine. It was a kind of dirty black and it had a rather fat little body perched too high on four spindly legs that suddenly spread out in panic as they approached the earth. A flimsy tray stuck in front of it like a black tongue, and on its side it had a most regrettable crank. Two wheels with S-shaped spokes (the artistic touch, no doubt) were held together by a fat leather belt. What looked like four hundred thousand little gadgets, spinners, springs, levers, gears, caps, screws, nuts, and bolts, were covered with a mysterious bluish down that looked like the mold on tired Gorgonzola. It was only a blend of paper dust and ink vapor. It was a sad machine, really, in spite of some gold striping that failed to lift its morale. Besides, it smelled. Smelled of oil, ink, and leather. This wasn't nice.

After looking at my patient for a while, under operating room floodlights, I decided that it was too far gone for a complete redesign job, as Angel Gestetner had allowed me only three days. A thorough engineering job might require from six to eight months in close touch with the development and production engineers of my client. And they were in England. So I decided to limit my efforts to amputation (the four legs) and plastic surgery on the body. By this I meant a face-lift job. I would simply encase all the gadgety organs of the machine within a neat, well-shaped, and easily removable shell. Then I would redesign the wheel, the crank, and the tray. The whole unit would then be placed over a set of four slender but sturdy legs, painted a pleasant color, and sent back into the business world.

The shell, concealing all the gadgets which were previously exposed, had other advantages. Being visible, they had to be finished carefully, nickel-plated, and polished by hand. A very costly process. Besides, the oily operational dust formed of pulverized ink, dust, and fine paper pulp would collect all over the machinery and interfere with proper functioning. It was very difficult to clean because of inaccessibility—and most unsightly.

So I proceeded to stack up plastic clay over the machine, and progressively I arrived at a form which enclosed everything that could be enclosed. I gradually improved that shape until I reached a form that seemed to me to be simple, practical, and attractive. I had made provisions for easy removability of the shell in simple sections to be hinged unobtrusively. After successive minor improvements, the duplicator was ready. It had taken three days, and it was enthusiastically received by my angel friend.

Sigmund crated it and shipped it, clay and all, back to his works in London. He has kept me on a retainer ever since, and built the same design ever since, too. It is an enormous success and even competitors admit that its appearance is still entirely up-to-date. This after twenty-two years.

Eventually, I was told that my theory had been confirmed. Thanks to the elimination of a great deal of hand polishing, the cost had been lowered considerably. Dust had been sealed out.

And it didn't smell any more. Well . . . less!

Industrial design was being born and I worked at it frantically. I felt that the ball was starting to roll. These were terribly exciting days. Soon I met Walter Chrysler, who took a liking to me. As he was a former locomotive engineer, we shared the same feelings for locomotives. I was on the verge of being retained by the Chrysler Corporation when a grand fellow named Jack Mitchell of Lennen and Mitchell, the advertising agency, beat him to it. I signed up with the Hupp Motor Company, a client of Lennen and Mitchell, and I believe it was the beginning of industrial design as a legitimate profession. For the first time a large corporation accepted the idea of getting outside design advice in the development of their products. And the fees were big-time, too. In this instance, eighty thousand dollars a year.

It is through my association with this company that I learned the facts of industrial design's life. The experience was such an ordeal that I nearly gave up the whole idea, thinking that it was all a mistake. I was almost convinced that American industry was not ready for it—and might not be for decades.

Jack Mitchell was a man of vision and intelligence. He was a cosmopolite who knew all about foreign motorcars, and he realized that the American automobile looked, as he used to say, like an abortion. He and I felt that something would have to be done quickly and that his client should jump in and take the lead. We both got very excited at the idea, which he sold to the company's president. So here we were all steamed up, and I got prepared for the great day. I took the Detroiter to Detroit, where a company driver waited for me, and we drove to the plant. Full of enthusiasm, I met the Chief Engineer in his office and he started telling me all the things I couldn't do. When it was over, about an hour later, it became apparent that the only things I was allowed to do were to jump out of the window, swallow a gallon of enamel, or sit under the ten-ton punch press. That fazed me a bit, but I began to recover.

Not for long, as I was to be given the Production Engineer's treatment. This distinguished gentleman wanted to clarify a few points that might not have been covered by the Chief Engineer. What he wished me to understand, see, was that whatever I would design that might pass by the Chief Engineer through some oversight, he was sure to catch anyhow.

Reeling and punch-drunk through this collapse of my dreams, I returned to New York and tried to collect myself. The following week I went back to Detroit for more and discovered a kind soul who felt sorry for me. J. N. was Body Engineer and a decent fellow. Unfortunately, he had some set ideas about automobile styling that were anchored tight and antiquated. However, I did not give up and J. N. and I progressively became quite friendly. Now, twenty years later, I realize what a nice guy he was, and how he really tried to help me. But let us see what I was up against.

At the time, cars were short, high, and stiff-looking. Windshields were vertical, there was a break in the hood, and windows were square. They looked more or less like this:

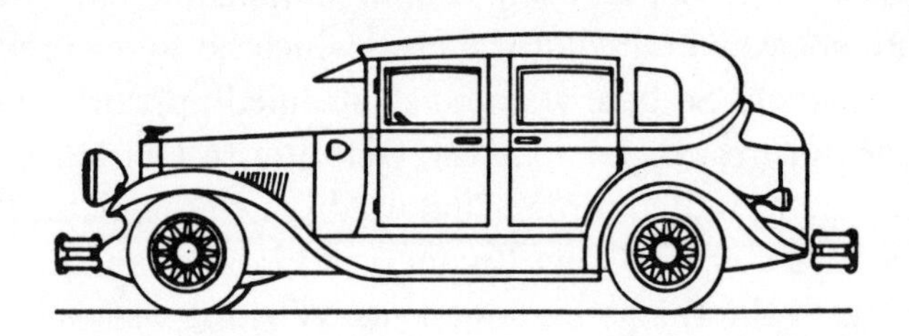

What I wanted was a car a bit longer, lower, and more graceful. It should have a slanted windshield, streamlined fenders, tapered rear, windows with rounded corners, etc. Something more or less like this:

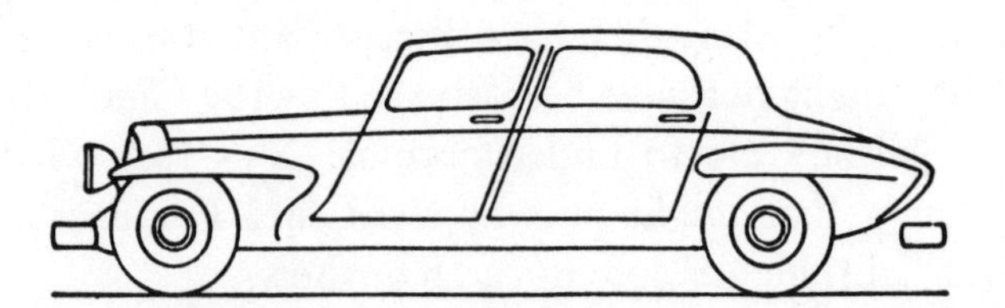

In other words, a car along modern lines, as shown on the design above, which I patented in 1930 and which still looks rather adequate.

When I had a chance to show my designs, I met a different kind of attitude. Surprisingly enough, there weren't any outspoken objections such as I had been led to expect. It was all very nice, evasive, and final. I wasn't told any more that it was not feasible; or too costly; or impractical. I was just told that my designs would be considered later, in due time, and that for the time being I might as well go back to New York until I received a call.

In other words, it was a clear case of the polite brush-off; but in my ignorance, I didn't realize it. So I kept on making sketches and talking my head off to anyone who would listen. After six months I had made absolutely no headway. A car had been designed without my help and it was considered for production. It looked awful: clumsy, top-heavy, and full of design vulgarities. It still had a straight windshield, rickety fenders, and a broken cowl line. I was discouraged and hurt, but I felt I couldn't give up. I knew I was right. I had to find a way to convince them, and it had to be done quickly.

Finally, I hit upon an idea. As they would not build anything of my design, I would build it myself, use it as an example. I knew that if they could see the design in the flesh, they would have to admit that I was right. So I went to work and in a few weeks an actual running model was built. In my hurry I had to pay for plenty of overtime, and it cost me eighteen thousand dollars. The car had some fresh features, slanted windshield, no cowl break, flowing rear end, etc., etc. With great reluctance, the top management finally agreed that it had many good points. It was decided that the model they had developed already would be delayed until I had a chance to prepare a full-size mock-up of my own design.

What a glorious day! I was leaping with joy and I forgot all my troubles and disappointments. Not for long. As the design was developed by draftsmen under the direction of the engineering department, I couldn't help notice that changes were made without my consent. Subtle changes at first, then more pronounced, and this over my objections. Soon the design lost its character and began to look more and more like the one that had been developed without me once before. What was most objectionable was the fact that it became taller, taller, and taller until it looked like a barn on wheels. Finally, I couldn't stand it any longer and I decided to talk to the key man. So I took him out for lunch to the Detroit Athletic Club. As we were driving along, I tried to convince him that cars were bound to get lower, not alone for appearance but for safety reasons, as it lowers the center of gravity. (They are now fifteen inches lower than they were at the time and they will get lower still.) Finally he said, "Loewy, you are all wrong and we won't go your way." We were stopped by a red light close to an old vintage trolley car. "You don't understand psychology," he said. "If we were now in one of your low cars, how would you like to be here and look up out the window at these people in the streetcar sitting way above you? You'd get an inferiority complex, that's what you'd get."

I realized then that I was up against something serious. It was a fundamental gap between him and me. Yet, I kept up my work and tried to minimize the damage. After many months of wrangling, many sleepless nights and days of worry and frustration, the car was produced. It had lost most of its dash and glamour. It was too high, too static and blunt-looking. However, there were sufficient new angles in it to give it a fresh look and it sold well.

Its introduction to the public coincided with the development of four-wheel brakes. The revolutionary new principle was explained to the young lady publicity director so she could prepare her releases. The new car, she was told, would have four-wheel brakes. The dear lady's engineering knowledge was rather thin (she used to run a tea shoppe), and she sent out a release containing this unforgettable gem:

> "This new automobile incorporates all the latest advances in the automobile world, including four wheels *and* brakes."

My frequent trips to the plant brought me into contact with the Detroit *Free Press*, a first-class newspaper, and with that national institution, Edgar Guest. I was reading his poetry with interest as I had been told that he was well liked and that it would help me understand the middle layer of America. The process, however, was not entirely painless as the poems often reached the fringe of infantility. One day, I told Tookie about my self-inflicted chore.

"Raymond, you should keep on reading it. It is good for you."

"But, Tookie, it is so hard to take sometimes."

"So, you agree with what the other poet said?"

"What did the other poet say?"

"Oh, you know:

> 'I'd rather flunk my Wassermann test,
> Than read a poem by Edgar Guest.' "

I kept on absorbing the wholesome stuff as I would have eaten yeast or yogurt, or done a daily dozen. It was good for me.

Chapter 7

SKYSCRAPER OFFICE

In the fall of 1933 I decided that I wanted more than one account. I was yearning to design plenty of different products, not just automobiles. So I took an office on the fifty-fourth floor of a skyscraper at 500 Fifth Avenue. I engaged two good designers, a secretary, and we got under way. It was a success from the start. The office was very up-to-date, simply and well decorated. The secretary was attractive, and the atmosphere pleasant. We had a repeating phonograph softly playing dance music all day long and everybody was happy. So happy that soon afterwards one of my two designers married the secretary.

I took a new and even more attractive secretary so my other designer could marry her. He did. I've had more secretaries married under me than most executives.

Soon we were doing so well that I was swamped with details and I decided to engage a business manager. It just happened that a young man I had met shortly before thought about the possibility of joining our organization at precisely the same time. He was an unusual fellow. Just out of one of the big three Eastern universities, he had been a top man on the football team. Let's call him Hugh Fenton. He was quite handsome, six feet five inches, weighed over two hundred and thirty pounds, and had the build of an athlete. His hands were like small hams. A socialite, Hugh was of independent means and certainly much better off than any one of us. His chief characteristic was that he was really a swell fellow in every sense of the word. The regulation stage Englishman would have said of Hugh, "The chap is a brick, reallah, y'know." Well, he *was* a nice guy and after a short while he developed a feeling of loyalty for me. I liked Hugh no end. He used to commute by air every day, landing his own plane at the Battery and flying back at night. The time came when Hugh felt he knew enough about the profession to get us some new business. Through some mutual friends he obtained an appointment with the Chairman of the Board of one of the largest corporations in the world whose office was on one of the upper floors of the Grand Central Building, straddling Park Avenue. On the day of the interview, Hugh had lunch at "21" with a couple of double Martinis to buck him up in the opening move of his new career.

Mr. D. received us correctly, motioned us to a chair across his desk—and it was Hugh's move. Mr. D. was a very, very small man, about five feet at the most, very frail and white-haired. Hugh, who had it all rehearsed, started his sales talk by saying, "Of course, *you* know Mr. Loewy?" No answer.

He repeated, "*You* know Mr. Loewy?" Mr. D. timidly waved

his white head in a negative manner. Hugh got purple. "Whaddya mean, don't you know him?" he said, pointing at me. Mr. D. was upset, and so was I. By now, Hugh had gotten up and he was towering over the poor dear Chairman, sunk in his chair and quite surprised. I was in a panic, too, trying to calm down my already former new business manager, who was ready to burst. I thought he might possibly do something regrettable and, with my imagination hard at work, I stared at the open window forty stories above Park Avenue. I could already see the headlines:

TYCOON DROPPED 40 STORIES
Thrown Out of Park Avenue Window
Industrial Designer's employee arrested.

I got up quickly, pulled at Hugh's sleeve, and managed to steer the conversation back to nonhomicidal grounds. Neither Mr. D.'s heart nor mine was in the conversation and the interview ended very soon. On our way down the elevator, Hugh kept on saying, "Cheesis, the S.O.B.!! Cheesis!! The S.O.B.! Cheesis!"

Hugh and I parted excellent friends and he has made quite a success since in ventures such as fighting in the Spanish Civil War, selling secondhand planes to the Turkish army, etc., etc. I often see him at the Colony, and it is always a pleasure. As for our prospect, it is one of the few large corporations which—in twenty years—we've never been able to sell, even remotely. I did not hire another business manager for years.

Among our first clients was a large textile manufacturer, the Shelton Looms, whose executives were imaginative and confident in the possibilities of industrial design. We were retained by them for many years. Another was an oil company for which we designed gasoline pumps and service stations. This kept us all quite busy and the situation looked promising; it might be that I had guessed right after all. Apparently, the urge to design products from a fresh viewpoint, this irresistible force which animated my organization, was felt by some members of another group of designers: the furniture designers. These gentlemen became restless and went to work. The results mark one of the saddest days in American "design" history. Having no inner spring of inspiration, little talent, and unfortunately poor taste, these fellows looked around for some peg on which to hang a different style of furniture. At the time, New York was trembling with the vibrations of thousands of riveting guns. Scores of enormous buildings were in the process of construction; a wonderful, dynamic spectacle that thrilled me no end. It was the golden era of the skyscraper. Simultaneously, the art galleries of Fifty-seventh Street, Madison and Lexington Avenues were sizzling with cubist art, an oily cataract of nudes descending staircases, or ascending them, all tangled up in a hash of geometric fragmentation. It was loud, ugly, and chic. Our stylish stylists of the furniture world, responding to the call of the cash register, put two and two together and created "skyscraper furniture." Whoopie! It was, as they used to say back in the 1940's, terrific. Then it all happened in quick succession. We saw the country flooded with one monstrosity after another: the Modernistic, the Cubistic, and the Futuristic.

My designers and myself were as staggered as the skyscraper furniture itself. We saw it all happen with consternation: the interior of the Chrysler Building, the Lexington Hotel, the Roxy, the Eighth Avenue cafeterias, the Paramount Theatre on Broadway, etc., etc. It was shocking and a sign of ill omen. If such nightmares of vulgarity were sweeping the country, what were the odds against us, the pure boys, apostles of simplicity and restraint? Our only hope was that some men of taste would revolt against these gilt outrages and turn to us for help.

Meanwhile, we tried to protect ourselves from getting swept away by this modernistic cataract as we kept plugging along.

CHAPTER 8

Some industrial design or advertising instructors like to tell students that one of the paramount requisites of any design (whether product or label or layout) is to draw the attention of the potential consumer. This statement should be elaborated. That goal is never to be achieved at the expense of logic, function or, simply, aesthetics. Throwing an egg into a revolving fan is certainly an effective way to attract attention. Merely to do the unusual for no particular reason is not sufficient. As an example, this page has been printed sidewise as an artificial device to attract your attention. However, it is very likely that you find this a bit silly; it may even have aroused a certain amount of annoyance and resentment. Its inclusion merely serves as a graphic illustration of the above theory in terms of typography.

AMERICAN COOKING

In 1929, I met a lovely young lady who was soon to become my wife: Jean Thomson, a beautiful Danish-American girl who had been brought up in a French convent. She spoke fluent French, we soon felt an irresistible attraction for each other, and we married in 1931. I had never been married before. We lived in a large apartment on Park Avenue where we kept Terry, a great big sentimental Irish setter. Whitney, the English butler, was a tall, ruddy fellow as formal as a stage butler and as well trained. Unfortunately, as soon as his duties were over, Whitney would get quite drunk, in a quiet and refined way. We did not mind it too very much as he was a lovable fellow and his service while sober was impeccable. Whitney adored Terry, our setter. One day Terry fell very sick. The vet came to see him and diagnosed distemper. He told us that Terry wouldn't last more than a few days. Whitney was heartbroken. He told us that he could save him if we would only let him take care of him in his own way.

We were so desperate that we agreed and Whitney carried the poor setter to his bedroom. For two days he wouldn't tell us anything and insisted that we must not go and see the dog. After three days, I couldn't wait any longer. I sent Whitney on an errand and stepped into his room. Terry was lying down in his basket. When he saw me, he managed to get up on his legs and staggered to my feet. I didn't think he looked too bad. In fact, he had a rather cheerful, if groggy, expression. I knelt down and patted his head. There was something funny about him. I opened his mouth and had one smell—Terry was reeking with Scotch. Terry was plastered! I put him back to bed, tiptoed out, and said nothing to Whitney. Jean and I laughed so much that we were just sick.

Late at night I went out on the balcony and watched Whitney take Terry out for his necessities. He carried him in his arms, reeling slightly under the weight. He put Terry down near a fire hydrant. Terry tried to lift one leg, faltered, and fell flat on his nose. Whitney was kneeling down next to him trying to steady him up. They finally made it and both staggered back to the hall, looking very happy indeed. A few days later, Terry was convalescing nicely, as well as ever.

Our friends the Joseph Platts lived in the same building. June, an epicure, is the author of the famous series of June Platt's Cook Books. She is also a beautiful young woman of charm and sensibility. Joe, her husband, is one of my most talented colleagues and a humorous fellow. We used to have cooking parties together, in our respective apartments, and it was great fun.

June's cooking was subtle and delicious. Mine too. June paid me the compliment of including in her book a recipe I gave her, and she used my name on the jacket as an endorsement.

Cooking parties have become popular since, and it is a mixed blessing. There is a certain kind of amateur cook that presents serious risks to your stomach lining. I refer to the man—usually it is a man—who prides himself on his culinary talents. His favorite concoction is a special sauce. His Sauce. The Sauce is a brownish goo that has simmered for hours and that contains, besides garlic: onions, tomato paste, prepared allspices, curry powder, barbecue mix, smoke extract, anchovy paste, mustard, olive oil, pimento, paprika, vinegar, and four kinds of prepared sauces.

These sauces, which the French call *sauces mécaniques* (or mechanical sauces), are usually manufactured by sauce corporations in twenty-thousand-gallon batches. They contain dozens of ingredients including mustard, turmeric, burnt sugar, poppy seed, cayenne, chili powder, cloves, tabasco, horseradish, soybeans, filé powder, meat stock, proteins, vinegar, pepper, pimento, salt, molasses, and benzoate of soda.

When The Sauce is poured over any kind of meat, fish, or fowl, the result is invariably foul. These amateur gentlemen never appreciate the finesse of a simple dish, simply prepared, where the particular flavor of the vehicle—whether meat, fish, or fowl—asserts itself, gently but firmly, with the gracious co-operation of a collaborating sauce.

American cooking is among the world's best when it is of a direct and straightforward nature. There is nothing anywhere to compare with a deftly broiled Maine lobster, or a couple of nicely grilled pork chops with apple rings sautéed in butter. Or a thick double sirloin steak, charred outside to a crisp over a charcoal brazier, nice and juicy on the inside; then sprinkled with crushed pepper, and a pat of fresh parsleyed butter melting over it. At its side (on a separate plate, of course), a majestic Idaho potato thoroughly blending itself with plenty of butter. A little crushed pepper and rock salt over it, but please, oh, please, no paprika.

As far as pies are concerned—and apple pie is the thing—they can be either fit for the gods or unfit for the dogs. There is nothing between. American cooking can also be one of the world's worst (leaving out, of course, English food, intended to be ingested solely for survival purposes). When the ordinary American cook attempts complicated recipes it is time for one to keep cool and to choose between a quick exit and sodium bicarbonate.

I retain bitter memories of certain "pan gravies," chlorotic béchamels, and deadly *sauces maisons*. Such preparations play into the hands of those people who question the validity of the American Century. They have no excuse and deserve no forgiveness.

What puzzles me is the fondness of the American people for blandness in food. It must be that they really like this sort of thing, because it sells in staggering volume, year in and year out. Is there anything more insipid than standard factory bread? So white, so fluffy, so soft, so dull. No taste, no flavor, no bread. Of course it is made of purest ingredients in prophylactic bakeries and it is vitaminized, homogenized, fluffed up, sweetened, enriched, milked up, slow-baked, triple-milled, cracked, sterilized, and "ultraviolated." As far as I am concerned, it might be as well made of talcum powder and distilled water. Seems a pity when I remember tasty French or Italian bread made of not too, too refined flour, with its crisp and sturdy crust. And what about that unforgivable mess, the processed cheese? Made with the best of ingredients, for me it is just a bland and vapid goo. Give me Camembert, Liederkranz, or even some good plain rat cheese. As far as the canned processed meat is concerned, it is a severe indictment of the nation's preference in nourishment. Is there anything duller than the vapid, pink, anatomical mess called luncheon meat? The quality of the ingredients in this case, too, is perfect, we know. But the result is mortally boring. No wonder I heard some caustic fellow in Europe call America "The Great Spamocracy, the land of the 'boiled baby.' "

For me, "luncheon meat" is, and will remain, boiled baby. George Nelson, one of America's talented young architects and writers, feels very strongly about the subject, which we have often discussed with passion and unanimity.

Then there is another nuance: the hostess who disguises food. Does it look like lobster? It is sweetbread. Does it look like sweetbread? It is apricot dumplings. The most vicious ones are the ones who make peanut butter appetizers look like goose liver, even faking the truffle with a thin disk of black olive.

Nathan Hale said, "I only regret that I have but one life to lose for my country," which, translated in free French, would read: "I only regret that I have but one stomach to lose for France." And they do, believe me.

One of my dearest friends in France is also the country's foremost epicure. I refer to Justin Laurens-Frings, now and for many years past President of the famous Club des Cent, one of the most exclusive clubs in France, whose membership never exceeds one hundred. All members are true gourmets and the requirements for admission, if ever written up, would make fascinating reading. By 1937 Laurens-Frings had never been to America. Like most Frenchmen, he was skeptical about American cuisine and hard to persuade that native cooking, of a certain kind, can be exquisite.

I thought that he could be best convinced by inviting him to my country home outside of Paris for a true American meal that I would prepare myself. It was a cool autumn day. When he arrived with his family, everything was ready. The décor was perfect. La Cense, my home, is a sixteenth-century hunting rendezvous built by France's great King, Henry the Fourth. The dining room was cozy and fragrant with the scent of acacia wood burning in the monumental fireplace, the dining table sparkling with silver and crystal, illuminated by the soft candlelight. There were autumn flowers in profusion, and the two Irish setters lazily sprawled on the henna carpet in front of the fireplace. The menu was thoroughly American:

Cream of Fresh Clams
Fried Baby Chicken Maryland with
Corn Fritters
Braised Endive
Romaine Salad
Old-Fashioned Strawberry Shortcake

The cream of clams was very light, not at all the heavy, thick type. The very small chickens, *previously boned*, lightly rubbed with garlic and dipped in batter, were fried very crisp and wrapped in napkins in order to remove any trace of excess fat. They were served very hot, in a tall mound surrounded by bouquets of crisp chervil and parsley. The chicken pyramid rested on folded napkins and the large platter was eighteenth-century sterling. The liaison of the sauce contained a very minimum of flour and naturally thick cream and chicken stock. A few drops of lemon and a generous sprinkling of crushed pepper gave it the right piquancy. The creamy sauce was brought to the table near the boiling point in a large silver boat with ladle.

The corn fritters were small, golden, and crisp, about the size of a fifty-cent piece, creamy and yet fluffy inside. Definitely on the side of lightness. The salad was well tossed, flavored with a trace of fresh tarragon and served chilled with sharp American Cheddar cheese.

The strawberry shortcake was a dream. The old-fashioned biscuit was covered with a generous amount of ripe strawberries at the last minute in order to avoid sogginess. The fruit had been crushed ever so slightly and allowed to remain for an hour or so in a light syrup to which it transferred its flavor and its adorable pinkness. A restrained amount of fluffy whipped cream was placed on top. An important point: the cake was oven-warmed, but the strawberries and whipped cream were cool. The contrast is pleasurable.

Then coffee and an iced raspberry brandy served in chilled snifter glasses. Laurens-Frings was convinced, to quote Pepys, that "my dinner was noble and enough." In fact, he has come to America three times since and plans to return for more.

My favorite poem in English literature (I know very little about Keats, Browning, or Donne) happens to be on the subject of food. Through the courtesy of the *New Yorker* magazine, I would like to reproduce it here:

THE STRANGE CASE OF MR. PALLISER'S PALATE*

Once there was a man named Mr. Palliser and he asked his wife, May I be a gourmet?
And she said, You sure may,
But she also said, if my kitchen is going to produce a Cordon Blue,
It won't be me, it will be you.
And he said, you mean Cordon Bleu?
And she said to never mind the pronunciation so long as it was him and not heu.
But he wasn't discouraged; he bought a white hat and the Cordon Bleu Cook Book and said, How about some Huitres en Robe de Chambre?
But she sniffed and said, Are you reading a cook book or Forever Ambre?
And he said, Well, if you prefer something more Anglo-Saxon,
Why suppose I whip up some tasty Filets de Sole Jackson
And she pretended not to hear, so he raised his voice and said, Could I please you with some Paupiettes de Veau à la Grecque or Cornets de Jambon Lucullus or perhaps some nice Moules à la Bordelaise.
And she said, Kindly lower your voice or the neighbors will think we are drunk and disordelaise,
And she said, Furthermore the whole idea of your cooking anything fit to eat is a farce. So what did Mr. Palliser do then?
Well, he offered her Oeufs Farcis Maison and Homard Farci St. Jacques and Tomate Farcis à la Bayonne and Aubergines Farcies Provençales, as well as Aubergines Farcies Italiennes,
And she said, Edward, kindly accompany me as usual to Hamburger Heaven and stop playing the fool,
And he looked in the book for one last suggestion and it suggested Croques Madame, so he did, and now he dines every evening on Crême de Concombres Glacés, Cotellettes de Volaille Vicomtesse, and Artichauds à la Barigoule.

OGDEN NASH

* Courtesy of the *New Yorker* magazine.

About food, too, are these two cartoons which I cannot resist inserting here:

"Oh, I like missionary, all right, but missionary doesn't like me."

"Look around! Do you see any *other* horses crying because *they* can't have roast duck?"

Reprinted from COLLIER'S

But let us return to the United States cuisine. It seems to me that the American woman may be partly responsible for the blandness discussed above. I've noticed that about the best food in America is not found in the tearoom shoppe, but at truck drivers' joints, near plants at workmen's eat shops, or Second Avenue bars and grills. There, you can always get a wholesome hamburger or some nice juicy pork chops with plenty of hot cottage fries, a hefty chunk (not a thin, patrician slice) of corned beef with real honest to goodness cabbage, and a glass of cool beer. I believe that if men were left alone they would soon demand real he-man's bread, not the sissified stuff that looks as if it were daintily made by some arty and desiccated spinster at Ye Olde Tea Roome Shoppe.

Okay, sister, you win. Bring on the pink candle and the lace paper doily.

Another surprising American culinary phenomenon is the Home Economist. These dangerous creatures try to keep the bored and frustrated hausfrau from falling into the hands of a psychoanalyst (which they couldn't afford anyhow) by keeping them busy in the kitchen. The formula is a blend of Poetry, Art, and Cookery. It gives the repressed, romantic mamma a chance to express her social amenities and relieve her libido through refuge in Arts and Cookies. A typical example of the home economist recipe ordinarily calls for an electric mixer and it runs something like this:

Fairies' Tails

QUEUX D'ANGES (IN FRENCH)

A Dainty Tid-Bit

SLICE TWO BANANAS LENGTHWISE, REMOVE THE SKIN AND PLACE IN A LARGE DISH ONE INCH APART. COVER WITH DAINTY STRIPS OF SMOKED SALMON PLACED CROSSWISE AND COVER WITH SHREDDED COCONUT. COAT WITH SWEET MAY-ONNAISE AND CHOPPED BEETS (FOR COLOR). COVER WITH DAINTY LACE PAPER DOILY AND LET IT REST FOR A WHILE. THEN DECORATE THE DOILY WITH CURLICUES MADE OF STRAWBERRY JELLY. POUR GENTLY ENTIRE CONTENT IN ELECTRIC MIXER AND BLEND AT HIGH SPEED FOR THREE MINUTES. REMOVE THE LOVELY PINK FOAM FROM THE BLENDER AND ARRANGE DAINTILY IN PINK SERVING DISH IN THE FORM OF CUTE LITTLE TAILS. SERVE COLD WITH A CHERRY ON TOP

Well, if it does the hausfrau any good—and I don't have to eat the stuff—that is quite all right.

It is funny how Art seems to be essential to the happiness of a lot of otherwise perfectly normal people who could pleasantly go through life just doing a good job and letting it go at that. But no, they've got to get all tangled up in Art. And here they go: arty cookies, painted lamp shades, Gauguin prints, and all.

A good example are photographers. In 1949, while being photographed in my apartment for a leading magazine, I received a frantic phone call from my housekeeper in Long Island. My home in Sands Point, two hundred years old, was on fire. I wanted to drive over at once, but the photographer insisted on taking his shots. I reluctantly agreed, frightfully upset. During that one last pause, he said to me, "Mr. Loewy, you look so serious." Here is a group of nice decent citizens who could load their cameras, stick the subject in front of it, push a button, make a print, send a bill, and call it a day. In fact, they used to do that for nearly a century, and they produced, without fuss, some of the most magnificent negatives of all times. But in the last twenty years or so, some photographers have become Arty. It started with the wearing of the hair long and the Svengali touch. They expressed their Artistic personality by trying to look like the Archbishop of Canterbury.

Some others preferred sloppy affectations, the baggy tweeds that are the badge of the high-brow (according to *Life* magazine). The pictures didn't get any better but the sweet gents began oozing Personality. Some are just dripping Art all over and they make Toscanini look like a lumberjack. Ready to burst with Art, they had to find a graphic expression of their aesthetic personality. Just like anything else, when Art's got to go, it's got to go! Luckily, the gents found an outlet: The Signature! The sluices were opened wide and they graced the world with an avalanche of Arty signatures that is here to stay and hard to take. This artistic tripe is typical of the average photographic art studio. On the other hand, there are men like John Rawlings, Irving Penn, Richard Avedon, Man Ray, and Cecil Beaton, men whose aesthetic contribution is in no way inferior to Modigliani's, Miró's, Picasso's or Stuempfig's. On page 112 are a few imaginary but representative examples of what one can see at the corner of many a commercial Studio Study in Sepia.

of beverly
«LENS ART
Artistry Studio
of hollywood.
fifth
avenu

Finely textured backgrounds either in product designs or in packaging have decorative possibilities as shown on the title page of this chapter.

Chapter 9

PENTHOUSE STUDIO

About 1935 our offices became too cramped and we moved to a penthouse office at Fifth Avenue and Forty-seventh Street. By that time we were retained by twelve firms or so and doing very well.

Doing well, but at what a price! The past four years had been a road hard to climb. No one in the manufacturing world had ever heard of industrial design, and no one was interested. My life was a dreary chain of calls on bored listeners. These men were doing well in their line and I was just a visiting nuisance to be brushed aside as soon as politely possible, or semipolitely possible. "Who is that fellow anyway, reeking with a foreign accent, a stranger in the land, to come here in my *own* office and try to tell *me* how to run my business? The nerve of some of those frogs [as they used to be called at the time]! Why, I've had the same design for twenty-two years, and nobody complains! Sorry, mister, I'm awfully busy right now. S'long."

Each link in the chain of calls was very much like the link to the right and the one to the left. I would like to be able to forget these business-getting trips in the Middle West, pushing doorbell after doorbell of small plants and factories. Whether in Chicago's Cicero, Pittsburgh suburbs, Toledo, or Cleveland, these trips were an ordeal. Will I ever forget those long treks in Chicago by smelly streetcars, rain-soaked, carrying a large brief case full of sketches, trying to get past a secretary to the assistant of some tough engineer or sales manager. And the long wait in the November rain for the sad streetcar that would bring me back to the heart of the Loop, to my hotel, tired and feeling grippy, disillusioned, lonely—and empty-handed. But above all, so tired.

Must stop at the drugstore off the lobby for a box of bromoquinine and some aspirin. Stuffy, flashy drugstore, crowded with members of the National Movers and Warehousers convention, each in search of a pint of bourbon and supply of life-savers, while Phyllis, the date, rouged and high-heeled, picks A Artistic Souvenir on the gift shelf—or a half ounce of Passion of Love. Then the crowded elevator, the reek of rye, the loud convention jokes, and, at last, *the corridor*. That long, ever so long corridor leading to 2745, just two doors short of the fire exit sign. That last mile past a thousand doors, transoms open, blaring radio broadcasts, drunken arguments, riotous laughter, girlish giggling, and, so often, a woman's sob. At last here is 2745. Home for tonight. Sad little room, so drab, so dull, reeking of cold cigar smoke and commercial orgies. It hasn't been cleaned well—one is so rushed these days. There are hairpins on the carpet, ashes in the trays, and Barbasol on the phone. The plate glass over the green commode is stained and sticky, but underneath it a gilt and crimson card attracts the attention of the guest to the Seven Different Cafés graciously placed at his disposition; seven in all. Now you can forget your deadly blues in the Dutch Kavern with an 85¢ blueplate special, including rolls, butter and coffee—dessert 25¢ extra. If you are flush the Management invites you to the Tuberose Room where Hal Halsted and his Saxoneers (broadcasting nightly over XZYK) will entrance you with "Tea for Two." Then Miriam and Feliz, the Society dancers, the sensation of Paris and London, will entertain you with an artistic rendition of Ravel's "Bolero" in the darkened room, under purple and green spotlights. (A welcome interlude for the now totally plastered conventioneer anxious to do a little exploring under the long tablecloth.)

Well, not for me tonight, thank you. Not in the mood. Instead, a nice long bath. Just time enough to remove an assortment of curly hairs from the tub and I'll be relaxing in the warm bath (chlorinated), with a copy of the *Evening Tribune*, reading all about the North Side slaying of Turp Tarazzo, the juke box guy. What a story. Turp was found in the A.M., drilled right through and propped up against one of his own juke boxes whose jammed mechanism had been playing "Chloe" all night long. Pork fat up ¼ cent. Boss Murphy won't run for Sheriff. Police Commissioner okays parking ban in Wabash. Tomorrow: Rain.

And next day, up at seven, and on my way again. I *must* try to get the Little Giant Dryer Company interested in correct styling. And so on and so on, for weeks, for months, for years, with some results once in a while—at ridiculously low fees. Take Sears Roebuck, for instance. It took me nearly two years of close-spaced visits to Chicago—at my expense, naturally—to convince them of the importance of correct product appearance.

I finally won, and I received the commission of designing the Coldspot refrigerator, at a fee of $2500. It cost me nearly three times that amount to do the job as I wanted it done. The design was accepted, built, and sales doubled. For the next model, my fee trebled. Sales increased enormously. My fee was raised to $25,000. Sales jumped to 160,000 units. Then to 275,000—a record, a thing unheard of in its field. Sears was convinced and they have placed the accent on design ever since. Nearly all of my early clients had to be "sold" in the same manner. The "show me first" attitude. Those years of pioneering were hard on me; they taxed my perseverance to the end.

As we shall see later, our work for the Pennsylvania Railroad started on a similar note of skeptical hesitation.

Yet I can hardly imagine anyone more suited than myself to design railway equipment. I have traveled hundreds of thousands of miles in all kinds of cars, from the early Pullmans of 1919 to the latest bar lounge. There was something curiously fascinating about the old green and red mahogany sleeper. The smell of "Fort Mackinowatok," in itself, was quite extraordinary. A melange of cold smoke, wet mohair, and fried potatoes with a dash of antique perspiration. But one got used to it very quickly, and good old lower 7 felt great after a long hard day at the plant. Unless the Machiavellian engineer chose to make some fancy starts that shook the dentures of some customers right out into the corridor.

The morning rendezvous in the washroom was a great spectacle—tall gents, fat gents, thin gents, all shaving away, pants sagging, suspenders swaying in unison, all in B.V.D. negligee, reeking with lilac toilet water, rushing to make room for the fellow next in line. I was always curious about the contents—widely exposed—of my copassengers' shaving kits. Some of these black or russet leather boxes, unswept for decades, were regular rats' nests full of incredible items. A Kinsey of the Rail will someday explore the American man's traveling kit.

In those days, the signature of an important new account was an occasion for great rejoicing. I would rush back to New York and announce the good news to a staff frantic with joy. We would get the champagne out of the cooler, and everyone, including office boys and charwomen, would drink a toast to the client, to the success of the design, and to ourselves.

Sometimes we would go to the German-American Club (in reality just a large and noisy German rathskeller) and have beer and sausages. Everyone would join in and sing—designers, division heads, secretaries, and office boys—and we would have a wonderful time until early morning. Why a German restaurant, I don't know. But it felt good, friendly, and informal. Everybody was happy, and it made me happy too.

Office boys have always played an important part in the atmosphere of the R. L. business setup. We seem to attract unusual boys. One became, in two years, assistant to our Director of Design. He had a brilliant war record, and now is at the head of a successful printing brokerage firm. His name is Dougherty. Several became crack designers and are making an excellent living. Some come to work in streamlined convertibles, and others walk to save the carfare. All have equal opportunities, and they take full advantage of it. They always do the unexpected. Recently, I wished to sell one of my cars before flying to Europe, an experimental car with a special body and rather expensive, being practically new. My secretary posted a notice on the bulletin board giving first call for the purchase of the car to the employees of the firm. Next day I found a check on my desk and a memo saying that the car had been sold to Jack Card. I did not know any designer of that name. It turned out that the fellow was an office boy.

They seem to have a certain sense of humor, particularly in the stockroom. On the walls they like to place mottoes such as:

FAMILIARITY BREEDS

NEVER BITE THE HAND THAT SIGNS THE CHECK

THEY SAID IT COULDN'T BE DONE
(So I didn't even try)

Imagination is a great part of our daily work, and I encourage all its manifestations. Our boys are aware of this and they do not hesitate to take up with the division heads any unusual idea they may think has possibilities.

Our attitude toward encouragement of imaginative thinking sometimes leads to surprising suggestions from our stockroom wizards. One came to me not long ago with a brainstorm. At that particular time, we had among our clients a large manufacturer of chewing gum and also one of the leading makers of toothpaste. Our boy had an idea for two new products destined for the large Italian population of the States: garlic-flavored chewing gum and garlic tooth paste. After all these years, I am still wondering, did we pass up a million?

On another occasion a fellow, probably annoyed to death by his four kid brothers and sisters, suggested the development of nembutal lollipops. The name? Lull-E-Pops.

Once we needed a good reproduction of a red flower called poinsettia, as visual reference for a Christmas package design. I asked a boy to go to the public library to get me photographs of poinsettias. After a while he came back and put on my desk some pictures of bird dogs.

"What's that?" I asked.

"Point-setters," he said.

When two new restrooms—men's and women's—were installed, the boys identified them with hand-painted signs:

HIS

HERS

My particular curve of business expansion, if studied by a seasoned and rational management analyst, would make his hair stand on end. For instance, in 1919, utterly broke, and when practically assured of a good position with the General Electric Company, I did not even go to talk to them, and launched myself instead into fashion illustration, a field of which I knew nothing, and a precarious one at best.

I succeeded to the tune of $40,000 a year.

In 1928, while successful and established as a fashion illustrator, I suddenly stopped everything to start industrial design, an unknown field, and by 1929 ran head on into the Wall Street crash that shook the world. It also shook $125,000 out of Raymond Loewy—practically all his savings. Most of my accounts were either sharply curtailed or canceled outright. I took one good look at the debris and decided that this was as good a time as any for expansion. So I rented a swank office on the fifty-fourth floor of 500 Fifth Avenue, furnished it sumptuously—and made a sparkling success of it until 1935. Then America was hit by the Great Depression. This was no time to get depressed. So I started looking for a luxurious penthouse office, found it on upper Fifth Avenue—and moved in. It did a great deal of good to our morale and we thrived in the place, expanding every few years.

In other words, regardless of what people may tell us, or experts may write, I stick to my basic philosophy. It is a simple one:

I believe that in our country there is always a chance of success for anyone

a) Who knows how to do a thing well
b) Delivers it on time
c) Sticks to his word.

On that basis, we went ahead successfully, depression or no depression. It is the sad privilege of the ineffectual, the lazy, or the imbecile to blame one's failure on the other fellow. Or to give up, if momentarily stopped. Or to sell short America's opportunities.

In the case of a corporation, long established, with a reputation for integrity and high business ethics, the trade-mark should express these basic qualities. The design below which Raymond Loewy Associates produced for the International Harvester Company seems to reach this goal in a forthright, simple manner.

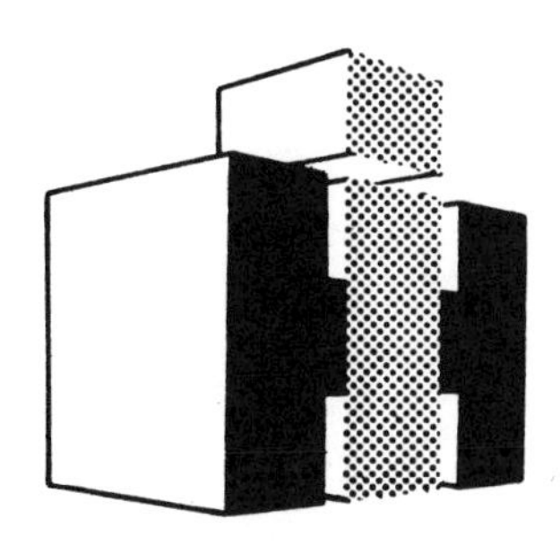

Chapter 10

THE "ME TOO" BOYS

The move to Fifth Avenue and Forty-seventh Street proved to be a good one. It gave a boost to our morale, already soaring, and we obtained more accounts. The sensational success of the Coldspot design was written up in various management magazines, sales and production periodicals, and eventually in technological publications. The case history of the Coldspot design became an active subject of discussion in merchandising and management circles. Numerous organizations and universities requested me to deliver talks, preferably illustrated, about our design technique. It was excellent publicity and it brought forth many inquiries from prospective clients.

This was one of the very first cases of styling for mass production by a design consultant and there was no precedent to go by. I made up my mind about a tentative approach, then my assistant and I started making rough sketches in small scale. Soon we discovered that such a sketch gives a different impression when "blown up" to actual size. More disturbing still was the fact that a full-size drawing would become even more distorted when built in three dimensions. We felt that it would be an advantage to work at once on full-size three-dimensional models, especially if we could use a material allowing for rapid changes and experimentations. Wood models wouldn't do. The obvious solution was clay. It had been very successful in developing automobile bodies. So we started making box frames slightly smaller than the finished refrigerator and we covered them with a layer of modeling clay, giving us a block of virgin plastic material to work on. Then we could carve or sculpt any shape we desired. It became apparent that the clay was easier to work with when warm, so we developed an electric clay-warming oven. When the models were ready for presentation to the client, we naturally wished to place them under the correct conditions of lighting and arranged in the right sequence. We discovered that they were too heavy to be moved without damage. So we built the next batch on low platforms mounted over roller bearings, permitting easy manipulation.

The hardware, such as hinges, latch, name plate, etc., was carved in wood and painted with aluminum lacquer. We felt it lacked sparkle and looked clumsy. So we sprayed it with a layer of copper paint. This metal finish was then polished carefully and the object placed in plating tanks. As a result we had parts that were perfectly nickel-plated and highly polished. The final object looked and felt exactly like a solid metal piece of hardware. All this at a fraction of the cost, and much faster.

When we started on our design, the Coldspot unit then on the market was ugly. It was an ill-proportioned squarish box "decorated" with a maze of moldings, panels, and other schmaltz. It was perched on spindly legs high off the ground and the latch was a pitiful piece of cheap hardware. We took care of all that in no time at all. The open space under the box was incorporated into the design and it became a storage compartment. The new latch was substantial and as well designed as if it had been intended to be the door handle of an expensive automobile, the hinges were made unobtrusive, and the name plate looked like a fine piece of jewelry. The entire connotation was one of high quality and simplicity.

The shelves had been a serious problem for the refrigerator industry. They had to be formed and assembled by hand, welded and dipped in rust-proof finish. They rusted just the same eventually, and the cost was exorbitant on account of the excessive amount of labor involved. Moreover they looked messy.

At the time, I was working in Detroit on the front end design of the Hupmobile. Among the materials considered for the radiator grille were several samples of perforated aluminum submitted by a supplier. In order to give strength to these large panels they were reinforced at the back by longitudinal ribs. The whole thing was extruded in one piece, very neat and quite inexpensive. While looking at a sample it struck me as being the perfect material for my Coldspot shelves. I ordered several samples with the right perforated slots, had them cut to the dimensions of the Coldspot shelves, and took them to Sears Roebuck. They were a sensation. Here was our answer, a simple one—attractive in appearance, foolproof in manufacture, and completely rust-resistant. They were adopted, imitated by the industry, and became the standard for many years. This was a windfall for the aluminum industry, which supplied the metal for several millions of shelves year after year.

Another improvement was the "feather touch" latch. This latch was designed so that a housewife with both hands full could still open the refrigerator by pressing slightly on a long bar with her elbow. In addition, it was connected by remote control to a small foot-operated pedal close to the floor. All these features combined made perfect advertising material for the copy boys and supplied the salesmen with great sales-talk features. Most important of all was the realization that our analysis of the problem, in close co-operation with our client and the engineering staff, had simplified not only the appearance of the product but also had lowered its cost. The results were spectacular and it made our reputation. It also put industrial design on the map as a fundamental new factor in merchandising. As I mentioned before, this case history became known and it was the subject of careful study.

No group of men studied it more carefully than my skyscraper-design friends. These gents could already hear the crash of the short-lived skyscraper furniture boom, and the most alert ones were looking for something else to do. It didn't take them long to discover industrial design. It was a stampede. Some of them, having persuasive ability, managed to get design commissions out of credulous manufacturers and they produced an unbelievable amount of imbecilic stuff. In most cases the designs were so impractical that their clients threw up their hands and called it a day. Some others actually found a way to build the contraptions and it wrecked, or nearly wrecked, their business.

Besides the serious designers such as Dreyfuss, Van Doren, Ray Patten, Arens, Teague, Sakier, Wright, and others, we had to contend with a group of twenty or thirty crackpot commercial artists, decorators, etc., without experience, taste, talent, or integrity, who called themselves industrial designers. They nearly killed in the nest the young profession that we were trying to nurture through adolescence. Recently S. M. Schweller, Chief Engineer of General Motors' Frigidaire Corporation, was telling me a story typical of these early days. A rather well-known so-called industrial designer used such high pressure about getting an interview that, against Frigidaire's best judgment, they finally gave in and set a date for a meeting at the plant. This, the designer said, was not satisfactory. In order to get the full value of his novel ideas, it would be necessary for them to go to a leading Dayton hotel for his presentation. So one evening, the executives dropped in at the Van Cleve and they were directed to a suite upstairs. There they were greeted by the Artist and led to a darkened room, candle-lit and mysterious. Violins were playing soft gypsy music. A beautiful model in décolleté gown was languorously draped against The Design set before black curtains. The icebox, in model form, was enormous, ugly and . . . *red*! My Frigidaire friends, among whom are some of the most brilliant men in American industry, were a bit staggered by the shocking vulgarity of the performance. They took one polite look at the design monstrosity and excused themselves.

This is the sort of stuff we had to fight against. So I felt that our organization's chance of lasting success was to remain businesslike, practical, and realistic. The results speak for themselves. Some of our clients, among the largest corporations in existence, have had us on a retainer for twelve years or more. Our record is fifteen years with the largest railroad company in the world. We have established for ourselves a reputation for reliability and soundness of judgment. They trust my partners, my designers, and they trust me.

The Sears Roebuck success was a turning point in my career. I became extremely busy and soon reached a design production ceiling. I had two choices: either to do all the designing myself and remain at the same level, or to select two or three of my best young designers and develop them into top-flight designer-executives. In other words, I had to decide whether or not I would remain a one-man organization or accept the idea of delegating authority and thus grow to be an important firm.

After careful analysis of the men I had with me and of the potentialities of the profession, I made my choice. And I took these immediate steps:

1. Promotion of three key men to executive positions.
2. Hiring of a first-class business manager.
3. Establishment of a publicity and public relations unit.

The results were everything I had hoped for. The quality of the work improved further and we gained many new clients. Approximately two years later I took complementary steps. The most important were these:

1. Addition to the staff of technicians and engineers.
2. Creation of a model shop.
3. Creation of a clay and plaster modeling department.

All these moves brought immediate dividends. My assistants were alert, talented, and good mixers. My clients liked to work with them, and I began to see possibilities of big business. Soon after, we doubled the number of our accounts. Throughout this stage of our growth, I considered my own duties as vital and clear-cut. I set a fundamental rule that everyone was compelled to follow without argumentation. That rule, which was unalterable and permanent, was that:

NOTHING IS TO COME OUT OF R. L. OFFICES UNTIL IT HAS BEEN CHECKED AND DOUBLE-CHECKED FOR PRACTICABILITY AND MANUFACTURABILITY. HEADS OF DIVISIONS WILL BE HELD DIRECTLY RESPONSIBLE FOR THE OBSERVANCE OF THIS DESIGN POLICY.

R. L.

So this was my design philosophy and its continued success has substantiated Santayana's statement that "it is a great advantage for a system of philosophy to be substantially true." My reasons for the rule above, which has been in constant operation ever since, are obvious when one takes into consideration what some of our fly-by-night competitors were doing at the time.

"THAT PACKET OF ASSORTED MISERIES WHICH WE CALL A SHIP."
(Kipling)

Things were going well businesswise and I was craving some relief from work. Ever since my boyhood days I had been fascinated by boats, and at last I could afford one, so I made up my mind.

The fateful day arrived when I bought one. I have owned fourteen since and they have been the curse of my life. For twenty years, boats, or rather boat people, have made me miserable. I despise anything and most especially anyone connected with this odious yachting racket, but it is habit-forming—possibly more so than dope. I've taken more punishment at the hands of drunken skippers, degenerate cooks, and gypyard operators than I thought humanly possible. I've been taken: swindled and robbed by the greatest collection of drunken, ill-mouthed jerks and derelicts that ever drifted around a swank yacht club waterfront; men who would charter their own mother if she could float. I've had a steward charge me sixteen dollars for mustard in one week. I've had a "captain" tell me cold-bloodedly that he dropped six hundred dollars' worth of silverware in the Hudson River while putting the boat out of commission. I've had a cook tell me ten minutes before my twelve guests arrived for dinner that there wouldn't be any dinner because the bottled gas tanks were empty (he had been instructed a week ahead of time). I have been charged twice for the same gas and three times for the same oil. And every time at such an exquisitely chosen moment that there was nothing to do other than commit murder, or suicide, or pay the bill like a good sucker and try to swallow my livid rage in silence.

Two days before he fought Tunney in New York in 1924, Georges Carpentier spent the afternoon on my boat, with Lily Daché, Jean Desprès, and a few other friends. Unknown to anyone aboard, I was having some first-rate trouble with the crew. When we reached some isolated spot in the Long Island Sound, the captain said that the crew wanted a 33 per cent increase in wages retroactive two weeks. He slyly hinted that there was a great possibility of engine trouble and he wasn't sure we could be back to New York on time as scheduled. They got the raise and the engines worked very smoothly. We returned ashore and I fired the whole bunch.

I made up my mind never again to own a boat if I could help it. I couldn't. So I have owned five others since. The latest, a day cruiser which I bought a month ago, is a beauty—cruises at 40 MPH—and I will ship it to France soon. We shall use it in summer, at our home on the Mediterranean.

Chapter 11

FROM TOOTHPICKS TO LOCOMOTIVES

Among the very few persons who early sensed the potentialities of I. D. and its eventual impact upon the American way of life was my dear friend Mary Reinhardt. Mary is one of the important contemporary living Americans. Gifted with a transcendent mind, a rare sense of aesthetics, a boundless imagination, she became interested in my work at the very beginning. A visit from Mary to our office was always an occasion. Everyone loved her; she was so enthusiastic, so cheerful. As we are not exactly a sleepy bunch ourselves, her presence in our midst transmuted the whole place into a kind of homemade cyclotron. The atmosphere vibrated with new ideas, new angles, a veritable bombardment. Mary, who happens to be a beautiful woman, was a lovely catalytic agent for us to have in our midst. For many years, she gave us the benefit of her enthusiasm and her graciousness.

Mary, now Mrs. Albert Lasker, has transferred her interest and her dynamism to her home life and to a great variety of organizations in the field of health. Her chief interest at the present time seems to be the American Cancer Society. In addition she spends much of her abundant energy in financial legislation for medical research.

The start of this fund ought to be recorded, and here is the story:

In 1944 Raymond Loewy Associates gave to Mary, at her suggestion, a check for $2400 made out to the American Cancer Society Inc. in New York. She sent it with a letter to Dr. Frank E. Adair, President of the American Cancer Society containing this paragraph:

Dear Doctor Adair:

This gift, which was solicited on the basis that the Cancer Society intends to create such a Research Fund and small Research Committee, is, I hope, the first of many that will come to the Society in the future. The important thing is that the framework be established promptly so that research efforts in the field of cancer may be co-ordinated and directed along definite channels rather than continue as it has been in the past to be scattered and disorganized. (If, for any reason, the Society does not wish to have such a Fund or Committee, the contribution should be returned to Raymond Loewy Associates.)

Sincerely yours,
(signed) Mary Lasker
(Mrs. Albert D. Lasker)

A few days later Dr. Adair answered in part as follows:

Dear Mrs. Lasker:

The Executive Committee established a Cancer Research Fund and therefore your solicitation of the funds represents the first for the Society. I think we set up the Cancer Research Fund along lines which will be approved by you.

Very cordially yours,
Frank E. Adair, M.D.

Since then the Cancer Society has raised over $16,550,000 for research in the cancer field. So in a variety of ways, including this, Mary really started something. The whole field of cancer research is in a different area from what it was five years ago.

Other early believers in my one-man crusade were two men who volunteered to help me: Charles and Stuart Symington. Stu, formerly Secretary for the Air Force and now head of the National Resources Board, is one of the best friends I ever had. The Symingtons gave me a letter of introduction to the President of the Pennsylvania Railroad, Mr. M. W. Clement. So I went to Philadelphia for the Great Meeting with the head of the world's largest railroad. The outcome of this, I felt, would be of momentous import. The conversation went somewhat like this:

"Well, young man, what can you do for this railroad? First, tell me what you did during the war."

I told him briefly about my war years and M. W. Clement seemed interested. He had been on the French front and spoke a few words of the language. Apparently, he kept a pleasant memory of the experience.

"Have you ever designed any railroad equipment?" he said in his stentorian voice.

"No, but I have been dreaming about it for the past twenty years."

"So you like being in America?"

"I couldn't live away from it."

"Well, all right. We don't have anything for you now, but we'll get in touch with you if something turns up."

This wouldn't do. "Mr. Clement, I would like to start right now. I know I can do valuable things for you and the railroad, and I would like a chance to prove it to you. Can't you find one single design problem to give me now, today?"

"What do you have in mind?"

"A locomotive."

Mr. Clement is a white-haired, handsome gentleman of great physical stature. He looked quite taken aback and puzzled. He said nothing and kept looking at the ceiling, leaning way back in his swivel chair, thumbs in the lapels of his black coat.

After a long while, he pressed a button under his desk. His secretary walked in.

"Ask Brown to come over," he said.

Then silence again, while he and I waited. Once in a while he would furtively glance at me, and I was trying to look calm in spite of my unbearable excitement. I could sense my whole future being decided in this moment; I was breathless.

Brown came in, or rather rushed in to report to the Executive.

"This Frenchman has been recommended to me by Symington. He has a good war record and he likes it here. He thinks he can be a designer. He insists on doing a design job for us so he can prove how good he is. Look here, Brown, the trash cans in the New York terminal are terrible. Why don't we have better ones? I want you to make arrangements for this Frenchman to have a look at our trash cans. Maybe he'll be able to do something about it." And turning to me, "Go to it," he said.

"Thanks for the opportunity, Mr. Clement," I said. "I will convince you and you will trust me."

I was leaving his office when a tall, stern, extremely thin, white-haired gentleman walked in and Mr. Clement called me. "Just a minute," he said, "come back here." I turned around and faced the two.

"Ever seen an Indian?" he said, pointing at the terrifyingly desiccated executive.

"No."

"Well, meet one, then. This is Colonel Young, one of our Vice Presidents, a *peau-rouge*, as you say in your country; here we call them redskins." The Colonel was not amused and did not crack a smile. But somehow it did not scare me. We shook hands and M. W. C. said, "C. D., this is our new trash can artist. You keep an eye on him." With this it seemed to me that I perceived a twinkle in his eyes. He took my hand and gave me a very direct handshake—a lingering, friendly one.

"Okay, Loewy. Go ahead and take it easy." Upon which I left.

During the train ride back to New York I was so excited I couldn't even sit down. I kept pacing the parlor car up and down until I made myself a nuisance to the other passengers, and I finally sat down, mad with joy.

I went to the Pennsylvania Station and spent three days looking over the trash can situation. For hours I watched passengers, commuters, loafers in their most intimate dealings with a railroad refuse receptacle. I learned plenty and I went back to my drawing board. After a few days I had several designs ready for submission to the management. I did not see M. W. C., but someone okayed one of my sketches and a few samples were built. They were placed at strategic points in the New York terminal and I watched them in operation. A few days later I was summoned back to Philadelphia and into M. W. C.'s office.

"How's the great trash can specialist today?" he said.

"Okay. But how's the trash can?"

Upon which he installed himself comfortably in his chair and started on a long dissertation about the French, the war, and a dozen other things including French cooking. But not a word about my cherished cans. Finally I couldn't stand the suspense any more and I said again, "Mr. Clement, how's the trash can?"

"Young man," he said, "in this railroad, we never discuss problems that are solved." Upon which he motioned to his secretary, whom, evidently, M. W. C. had buzzed. "Tell Hankins to come over and to bring along a picture of the GGI," he said.

Then silence; M. W. C. leaning back in his chair, a repetition of our first interview. Hankins * walked in.

"Meet Loewy, our trash can wizard."

Hankins looked at me from under his bushy eyebrows and said and did nothing. He looked antagonistic and bored.

"Show him that picture," said Clement.

The photograph represented the prototype of the giant new electric locomotives, the famous GGI familiar to all those who travel between New York and Washington.

I took one good look at the photograph, trying to act calm in spite of my excitement. Then Clement said to me:

"This is a good locomotive. We just built it as an experiment. We are going to build more. See anything wrong with it?" I saw plenty but I was on the spot. I could hardly criticize it too harshly in front of the man who probably had developed it, and yet it did not look good. It had a disconnected look; component parts did not seem to blend together, and its steel shell was a

* Mr. F. W. Hankins, Chief of Motive Power and a man of extraordinary judgment and ability.

patchwork of riveted sections. It looked unfinished and clumsy.

"It looks powerful and rugged," I said, "yet, I believe it can be further improved, and I would like to borrow this print for a few days."

"Can he have it?" said M. W. C. to Hankins.

Hankins nodded reluctantly.

"Go ahead, Loewy. And take it easy."

"Mr. Hankins," I said, "I may have to ask some technical questions of your engineering staff. Whom should I see?"

"Me."

I left in a daze, overjoyed and thankful to these men, great leaders and engineers who had enough confidence to give me their time and such an opportunity to prove myself.

I rushed back to the office and announced the good news. We were delirious. I had already studied the photograph during the train ride back to New York, and it became increasingly clear that a grand job could be done.

First of all, I was thinking in terms of simplification. I wanted to show these men that I was no long-hair artist trying to pretty up a 6000 HP locomotive, but a realistic designer with practical sense. The first step would be to suggest welding of the shell instead of riveting. This would eliminate tens of thousands of rivets, simplify the appearance, and lower the manufacturing cost. Then we would work on details and accents.

I went back to Philadelphia to ask technical questions without divulging my aims. After checking and double-checking every angle for practicality we were ready. We completed our design in a few days, a magnificent airbrush rendering was prepared, and the great meeting took place with M. W. C. and Hankins.

The design caused a sensation—and a shock. A welded locomotive? What's the idea! Does he think he is designing automobiles? Really! The nerve of some people!

I lay low and hoped for the best, trying to weather the silent hurricane. My heart was in my mouth. The meeting was short and I returned to New York not knowing what to think. Then no news. I went back to Philadelphia. Nothing for me. More trips, and nothing but a vacuum. Dejected, I dropped in at the engineering department to say hello to one of the draftsmen who had helped me. Lightning seemed to strike me! What did I see on the board but *my* locomotive being drafted!

Fred Hankins saw me. "We are going to build a full-size mock-up of your design. In a couple of weeks I want you to come with me to Wilmington and we will look at it together, Loewy. And take it easy."

In my excitement, the one thing beyond my capacity was to take it easy.

We went to Wilmington. The locomotive, ninety feet long, was magnificent. Hankins was delighted. All the workmen liked it, always a good sign, and in spite of my efforts I found little to criticize. However, a few details were to be corrected: a radius here, a highlight there, a window corner, a recessed handlebar, etc. I had brought along a large roll of white adhesive tape and colored chalk.

"Mr. Hankins," I said, "I would like to make a few suggestions and it would be best if I could do it right on the mockup instead of on a sketch. Do you mind if I go to work right now?"

"Go ahead."

I took off my overcoat and with the help of a crew and some

tall ladders I began to make corrections. I knew exactly what I wanted, and did it with calm and precision. There was no hesitation on my part and Hankins, who was watching me, looked deeply interested. Groups of men were observing me perched on top of the mock-up, stretching tape, making chalk marks, and writing instructions on the surface of the shell. I could feel that my audience was convinced. The corrections were obvious improvements and all realized that I knew what I was doing.

We returned to Philadelphia. Hankins was friendly and happy, and I went on to New York. Eventually fifty-seven of these locomotives were built at a cost of eighteen million dollars.

Years later, it turned out that the welding process had saved millions of dollars. Besides, maintenance of the locomotives has been made easier on account of the smooth surface resulting from the absence of rivets. The welding technique is now universally adopted.

One thing led to another until we were called almost daily for consultation on some new problem. The range of these was unpredictable and always interesting. It could be anything: the color of a ferryboat, the design of a menu, a new signal tower, a bridge over the Potomac, a coffee cup, or the design of a bronze tablet for a retiring executive. Invariably, we tried to be imaginative and practical. Among the most interesting design commissions we received were the city planning of certain areas in a large town on the Eastern seaboard, and complete blue-ribbon trains such as the "Spirit of St. Louis," the "Admiral," the "Broadway Limited," etc. We even designed toothpick wrappers.

Some type faces are so beautifully designed that individual letters or ciphers have inherent decorative value alone. When this beauty is intensified by introducing the element of contrast (i.e. large size and black against white in this case), the result is forceful yet free from vulgarity. This device, sometimes used in product design, is very successful on labels.

Chapter 12

BIG BUSINESS

One autumn day in 1940, a well-dressed, stocky, middle-aged gentleman of ruddy complexion walked into our office unannounced, accompanied by his chauffeur. "I am Mr. Hill," he said, "and I would like to see Mr. Loewy." My secretary, Miss Peters, showed him in. He walked in silently, looked at me carefully, and said, "I am George Washington Hill and I want to talk to you." Upon which he calmly removed his tweed jacket, displaying a magnificent pair of embroidered suspenders. Turning to my secretary, he said, "Do you mind, ma'am, if I keep my hat on?" Beautiful Helen Peters smiled back. I was fascinated by the felt hat, a battered affair, the rim of which was pierced by an assortment of trout-fishing hooks and flies.

"My friend Albert Lasker tells me," he said, "that you don't like the Lucky Strike package. He also tells me that you 'think' you could design a better one. I don't believe it. Besides, there is nothing wrong with the package."

I remained silent so as to politely emphasize the contrast between this negative statement and the fact that he was sitting opposite my desk. He looked at me for a while, then smiled, said "*Très bien,*" and we were friends. Without further ceremony, he pulled out of his pocket a lovely cigarette holder and other accessories made of smoked bamboo and yellow gold. "Cartier," he said. "Only the French could make this." I thought the things were really in good taste. Then, a lighter, also well designed.

"And look at these suspenders! Cartier too."

"So are these," I said, showing my own, which they had made for me.

"Well," he said, "what about that package? Do you really believe you could improve on it?"

"I bet I could."

"Bet what?"

"Fifty thousand dollars." He remained silent an instant, looking out the French doors at the flowers on the terrace. He then took a memo pad from the polished desk, made a note in pencil, and threw it in front of me, where it spun like a roulette wheel and slowly came to a halt, right side up. It read:

LUCKY STRIKE package: March 14, 1940
$20,000 retainer

If we use result
$30,000 more.
G. W. H.

"How's that?"

"Very nice, Mr. Hill. Thanks."

He then got up, put his jacket on, walked slowly in the direction of the door, bowed to Helen Peters, and turned to me.

THE EARLY DAYS OF THE MACHINE AGE

MESSY

DIRTY

NOISY

BULKY

THE GESTETNER DUPLICATOR

The Author's Very First Design Problem—1929

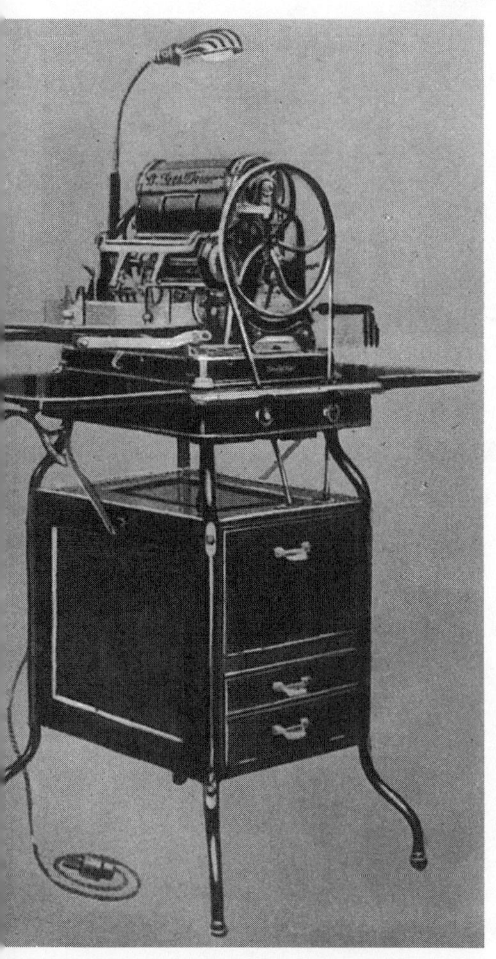

Prototype of the 6,000 hp, 120 m.p.h. Pennsylvania Railroad S-1 locomotive.

Opening of the pheasant-shooting season at La Cense, with Irish setter Gloucester (usually called "Glossy").

GREAT DAYS IN THE LIFE OF AN INDUSTRIAL DESIGNER

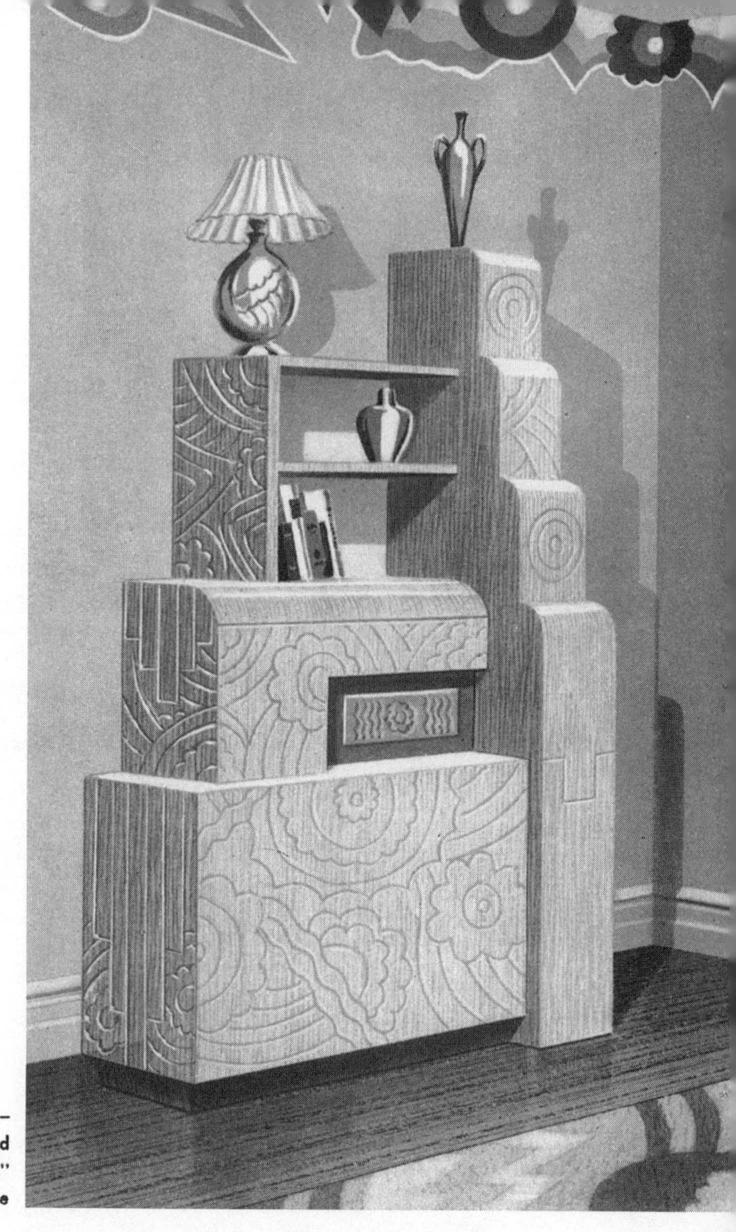

SKYSCRAPER FURNITURE—
"... ornament is but the guiled shore to a most dangerous sea."
—Shakespeare

SEARS ROEBUCK'S COLDSPOT

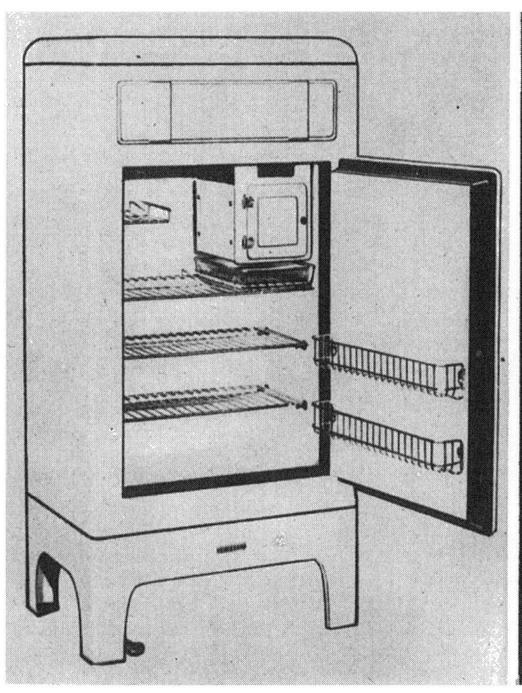

BEFORE–Sales: 60,000 units

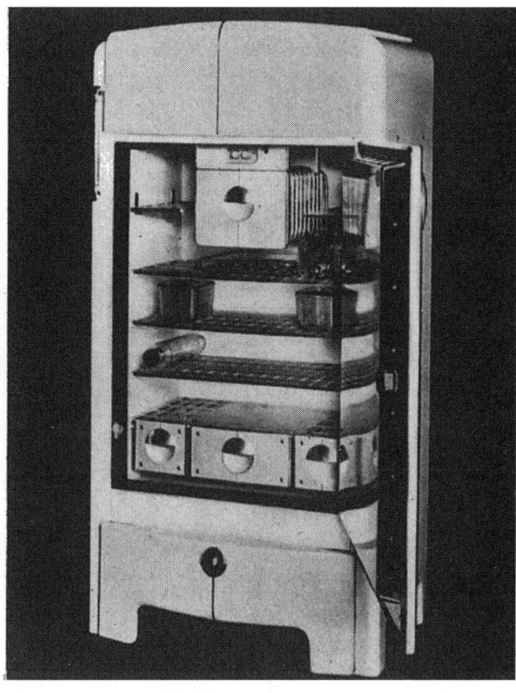

AFTER–Sales: 275,000 units

THE PENNSYLVANIA RAILROAD'S

BEFORE–Ventilating louvers are conspicuous. They break up the continuity of the design. Body is riveted.

AFTER–Louvers are scarcely visible, having been absorbed and blended into the horizontal effect. Body is welded.

GG-1 ELECTRIC LOCOMOTIVE

The prefabricated, welded shell is lowered over the chassis exactly as in the case of a Ford or a Chevvy.

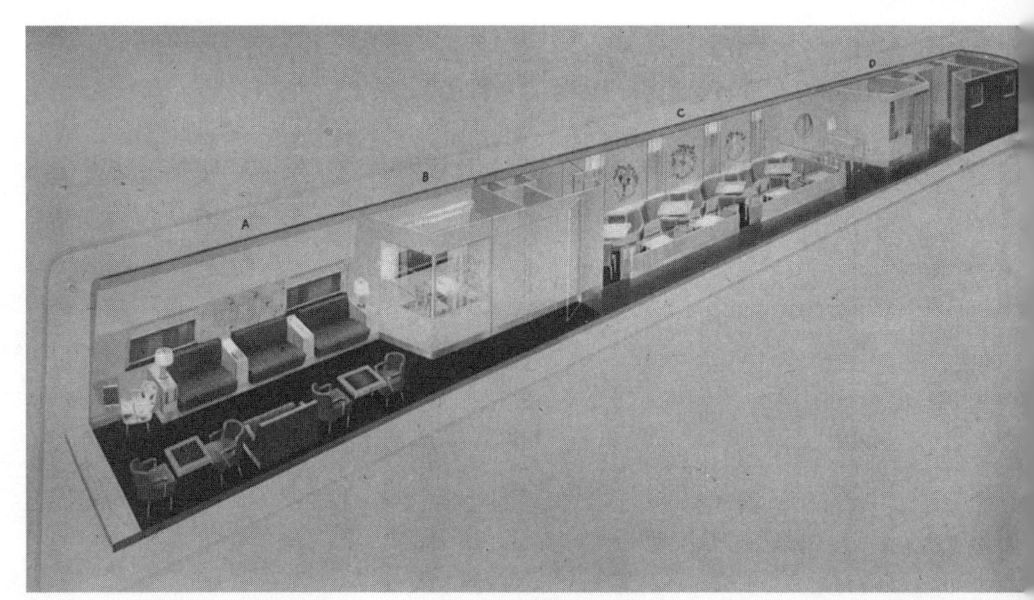

A—Games Lounge. B—Nursery, with nurse in attendance. C—Bar Lounge. D—Newsreel Theater. Planned and designed by Raymond Loewy Associates.

Bar Lounge. ***Gottscho-Schleisner Photo.***

RECREATION LOUNGE CAR

Nursery. ***Gottscho-Schleisner Photo.***

Games Lounge. ***Gottscho-Schleisner Photo.***

S-1 Prototype—1937

PENNSYLVANIA RAILROAD HIGH-

T-1 Prototype—1939

SPEED PASSENGER LOCOMOTIVES

BEFORE–The old Lucky Strike package was dark green. On the obverse was the well-known Lucky Strike red target. The reverse was covered with text that few people read. The green ink was expensive, had a slight smell.

AFTER–The new package is white and the red target has remained unchanged. The text on the reverse has been moved to the sides, displaying the red target on both faces. Printing cost has been reduced.

"When will you be ready?" he said.

"Oh, I don't know, some nice spring morning I will feel like designing the Lucky package and you'll have it in a matter of hours. I'll call you then."

We shook hands and George Washington Hill left with his chauffeur.

About five o'clock that evening the chauffeur returned with a package. It contained a carton of Luckies and George Washington Hill's card with a note penned in:

> "Put these under your pillow, and pleasant dreams.
>
> G. W. H."

It worked as I had anticipated. One day in April, I felt like working on the problem. I made a few sketches which satisfied me and we prepared a half dozen dummies. I called G. W. H. on the phone and he said, "I'll be right over."

After removing his tweed coat, he took one good look at the dummies; I had won my "bet." However, he had a few suggestions of his own which he wanted to try out. I had some colored paper, pen, ink, brushes, scissors, and rubber cement sent up and I placed everything in front of him. He started cutting strips of paper, red targets, gold or green bands, etc., and tried all kinds of variations. This for hours on end without saying a word. Mr. Hill was completely happy. I would leave him alone, at my desk, and continue my work, dictation, etc., with Helen, my secretary.

Finally, he got up, contemplated his work thoughtfully, pushed everything away, and said, "All right, Loewy, you do it," and left, contented and at peace with the world.

After all these years, after all I have read about this legendary figure, his alleged ruthlessness, violence, and business cruelty, I remain unconvinced. I retain of George Washington Hill the memory of a kindly man of superior intelligence and imagination. A man of sensibility, finesse, and savoir-faire. I know how unpopular this statement will make me with countless people, but I stick to it.

For those unfamiliar with package design and sales psychology, a few clarifications about the Lucky Strike design problem are in order. It has been proved on many occasions that a change in the appearance of an accepted product thoroughly identifiable by the public is a risky thing. If correctly done, it usually brings results immediate and lasting. It conveys the feeling of liveliness, of freshness. Above all, it must not destroy the identity of the package established at the cost of hundreds of millions of dollars, as in the case of Luckies. Any error can have serious consequences. Second, the design change must be a progressive one. Two intermediate steps or more are advisable before the final appearance is reached.

In this case, some definite improvements were achieved. In the original package, the famous red spot (called the target) appeared on one side of the package only. The other side was used for factory identification, federal regulations, inscriptions, etc.—items of no interest. All this text I transferred to the sides of the package, liberating the back area for a second red target. Thus, a package lying flat on a desk, a wrapper discarded, would inevitably display the brand name. Since it was designed, over fifty billion packs have been sold. Therefore, the Lucky Strike target has been displayed twenty-five billion additional times at no advertising cost to the American Tobacco Company. Besides, owing to its impeccable whiteness, the Lucky pack looks, and is, clean. It automatically connotes freshness of content and immaculate manufacturing. The sales results confirmed the logic of the move.

I can see the day when all sorts of nationally known packages will go through a periodical freshening-up process of advantage to both manufacturer and consumer.

And designer.

In September, 1939, some of my thoughtful friends unexpectedly congratulated me on the occasion of my twentieth year in America. It was both surprising and startling. Sometimes, under a shock, one realizes the existence of things otherwise taken for granted. I had been so busy forging ahead, so intent on trying to build something, that I had had no time for looking back at my own life. Like a parachute flare over a battlefield, war suddenly brought everything into sharp relief. Twenty years had passed since I had landed in America. Where did I stand? What had I achieved? Where was I going?

My original one-year trial of industrial design had gone past totally unnoticed in the rush of the decade. Instead, I was at the head of a relatively important firm. We had dozens of clients, most of them major corporations. In my office more than a hundred people were making a good living steadily, year after year.

We were becoming well known and our reputation, both ethical and professional, was impeccable. From the hundreds of suppliers, selling hundreds of millions of dollars of materials or products to our clients—often on our recommendation—we had never once been approached, even deviously, with offers of compensation or commissions. Our reputation for integrity was such that a move of that order was unthinkable to any supplier. We had gained the respect of everyone we had worked for. This was an achievement I had never fully realized before, and it made me proud.

Our expansion formula made it possible to create branch offices. At present, four are in operation, besides headquarters in New York. They are in Chicago, Los Angeles, South Bend, and London. They successfully operate on the basis of delegation of authority to top-flight design and business executives and intimate contact with the partners in the home office.

Among other rules of major importance are these:

1. Delivery of designs on time.
2. Careful follow-up with the client's engineering staff.
3. Constant checkup of the client's competition.

War naturally affected our operations considerably. The draft took away about 50 per cent of our men and we had to manage on a reduced staff. We developed new designers through accelerated training and re-established the balance in short order. At the very inception of my friend General William Donovan's OCI (Office of the Co-ordinator of Information) I went to Washington and lent to him one of my most brilliant young men. When OCI was transformed into the OSS (Office of Strategic Services), my "boy" became one of the top men in the department of Visual Presentation, possibly one of the most secret and important components in the strategical planning of the war. Incidentally, our organization, together with Teague's and Dreyfuss's, collaborated on the design of the layout and technical equipment of the office of General Marshall's headquarters. Throughout the war, we worked in the development of new ideas and new equipment for the Army Medical Corps, the Corps of Engineers, the Ordnance Department, and the Quartermaster General. I also had the pleasant opportunity to be in close contact with that great crusader and lucid fellow, Major Alexander de Seversky.

In 1944, as an expression of confidence and gratitude to my key men, I took them as partners and we became Raymond Loewy Associates. It gave further élan to the organization. By that time we had added a large area of drafting-room space below the penthouse, which in turn became the executive, accounting, and public relations offices. At present our operations consist of four divisions: Product Design, Transportation, Packaging, and Specialized Architecture. This last division is operating under the name of the Raymond Loewy Corporation, a closed-stock company, formed in May, 1949, and consisting of the same executives as the partnerships.*

My partners are in their late thirties and early forties—Americans, and married. All of them have been with the firm for ten years or more. Jack Breen, our Irishman with the Cheshire-cat smile, is the business manager. With him, the business is managed, and stays managed. No doubt about it. One of my client friends, impressed by his cold-blooded approach to a business deal, said to me, "Breen! This fellow has Freezone in his veins!" He acts as the flywheel of the organization.

A. Baker Barnhart, in charge of the products, automotive, and

* A. B. Barnhart, Jean Bienfait, J. B. Breen, W. T. Snaith, and Raymond Loewy.

packaging divisions, has been with me longest. A fellow of taste, he is not only talented but everybody likes him. Whether men on our own staff, clients, or suppliers, everybody agrees that Barney is a grand guy. Sensitive and keen, he is at his best in complex situations where the human angle is foremost as a factor.

William T. Snaith is one of the most remarkable brains that I have ever met, in or out of business. Some people who know him well suspect that he is not unaware of his brilliance. Bill has absolutely no design inhibitions. His radically new conceptions on department store layouts and efficient operations, his tasteful approach to decoration, have admittedly affected store retailing technique throughout the continent. His ideas are already permeating other countries. A brilliant partner, Bill is fundamentally an artist, and he has the characteristics of the artist. He is either in the dumps (I mean the real low-down, abysmal dumps) or soaring into the stellar space like an hysterical rocket. This characteristic has been known, at times, to be a bit on the trying side, but one gets used to these cycles, the recurrence of which can be pretty well charted.

Everything Bill does, he does well. As a sailor he is good. As a painter he is exceptionally good. One of his canvases was shown at the Whitney Museum in New York. A painting of his, "The Tower of Babel," hangs in my home surrounded by Picassos, Mirós, Dufys, Augustus Johns, etc. Many a connoisseur of modern art tells me (and I agree) that Snaith's work compares very favorably with the masters'.

Jack, Barney, and Bill are men of strong personality. I singled them out years ago, when they were unknown and making a modest living. I have often congratulated myself on the sense of human values that prompted my choice. It had a deep influence on our successful business evolution. Hard workers, cheerful men, they combine imagination with a sense of practicality. But the one saving grace, the common denominator of their talents, is enthusiasm.

There are other shining key men in the R. L. A. setup. Franz Wagner, for instance, the manager of our Chicago branch and an architect by training. Franz has a flair for organization and the Chicago branch is a model of efficient management. Its business curve is a gentle, lovely upward sweep, a pleasure to the eye.

Clare Hodgman, one of the country's foremost designers, is head of our products division. Gifted with a delightful sense of icy humor, he is the ideal traveling companion for those long stretches through the Middle West.

Herb Barnhart (no relation to A. B. Barnhart) is a crack engineer and design executive. Among the other stars in our own private constellation are Harry Neafie (transportation design division), Maury Kley (department store division), and Allmon Fordyce (specialized architecture).

There are many others whom space will not allow me to mention in detail: the Lathams, the Morgans, the Larsens, the Hunts. All contribute their share to maintaining the breathless pace of our organization. All of them are fun to work with, all are gentlemen.

My private secretary, Miss Helen Peters, who has been with us for thirteen years, is exceptional in many respects. Miss Peters (known throughout the organization as Pussycat Peters) is very beautiful, young, and witty. Blond and blessed with a perfect figure, usually dressed in a neat black dress with white piqué cuffs and collar, a strand of pearls accenting her graceful neckline, she is delightful to look at. As a personal secretary, she is the executive's dream. Intelligent, fast, and accurate, she is a definite asset to the firm. I owe Helen Peters my thanks and my gratitude for all the wise and timely decisions she invariably makes, spontaneously, without instructions, and for the gracious way in which she handles ticklish situations. Our clients love her and appreciate her. It is the sad duty of Helen, poor kid, to get theatre tickets for clients coming to town unexpectedly. As they invariably want to see the main hit, it often drives us to near insanity.

Helen's eyelashes are three-eighths of an inch long.

So by 1945 we were in full swing, retained by over seventy-five corporations. Our dealings were for the greatest part with top executives of the largest corporations in the land. They called us in consultation whenever they had a problem of importance. We were big business in our specialized field.

About that time an incident happened that influenced considerably my philosophy of business life and of social responsibility. It occurred in a natural way and it left an indelible mark. I was on a visit to a client in Dayton. The product we had designed, Frigidaire refrigerators, was an outstanding success; sales went sky-high. The company, the largest in the world and the leader in its field, had increased its lead and we were all happy. My friend Mr. Biechler, the General Manager, asked for me.

"What are you doing tonight, Loewy?"

"Nothing special, early dinner, probably, and a movie."

"Why don't you come to my home and have dinner with us? I'd like to show you something."

"I'd be delighted."

I was mildly puzzled about what he wished me to see, and I arrived at his home. A comfortable, luxurious house, his wife a charming hostess, a cozy fire, deep chairs, and a magnificent bottle of vintage champagne rosé. We had an excellent dinner and I felt relaxed and happy. I had all but forgotten that my host had expressed a desire to "show me something" when I noticed that he was looking at his watch at frequent intervals. We were enjoying a remarkable old Armagnac when the butler announced that the car was ready.

"Come on, Loewy, let's take a little ride together."

I said good night to my hostess, and we left. It was one of those perfect Ohio winter nights; the air limpid, myriads of stars sparkling in the sky. We drove in silence through the countryside. In the distance we could see the highway, a continuous ribbon of lights moving swiftly in opposite directions. It was shift-time at my friend's gigantic plant, and the road was covered with literally thousands of automobiles. We reached the four-lane highway and blended in the stream going plantward. The pace was even, the flow regular, silent but for a rhythmic beat as we passed each car in the home-bent lane. No sounding of horns, no brake screeches, only a mighty purr, a feeling—of order, precision, power. As we reached the crest of a hill, we could see the stream of red taillights and the stream of white headlights fading away in the distance. The sprawling plant was ablaze with blue mercury light. Over certain areas, the sky was shivering with the blue-white flashes of automatic welding. White, red, green, and blue signal lights would punctuate the night. The whole sky was aglow.

My friend leaned over, nudged me gently, and said, "Ain't it purrty?"

Pretty! What a word; what a subtle guy! I was utterly moved by the magnificence of it all. It was like seeing the actual flow of the rich red blood of young, vibrant America.

We reached the plant, entered the gates, and, without stopping at the executive offices, went right through the assembly line. It was action at its rhythmic best, a furious tempo on the basso side. Speed without confusion. The product could be seen, gleaming white, moving on the chain far into the distance.

We paused in a quiet spot, and Biechler took my arm and said:

"Loewy, my friend, I wanted you to look at all this. You see, when you and your boys work on our problem, in your penthouse office on Fifth Avenue, you may not realize the real importance of the pretty lines you put on paper. You see, every one of these men around us supports a family of four (as an average). Think of the wife, mom, or the old man, the kids. They all live well, because they have a job. They have a job because, among other things, your design clicked. In this plant alone—and we have dozens of others all over the world—eighteen thousand men are employed. Eighty thousand dependents! And remember that for each man employed at the plant, there are three in the field: salesmen, advertising men, maintenance men, traffic and transportation fellows, warehousers and accountants, dispatchers and repair crews, electricians, statisticians, engineers, draftsmen, etc. That's another sixty thousand. If you add to that another two hundred and fifty thousand for dependents, you get a true picture. More than three hundred and twenty thousand people whose life is directly affected by the success or failure of what you put on paper."

It made me thoughtful, to say the least. My friend motioned to a tall, white-overalled foreman. "Parks," he said, "I want you to meet Raymond Loewy, who designed the unit for us." Parks smiled, removed his glove, and we shook hands. "Thanks for a good job," he said.

I often tell this story to my employees, especially to new designers who join the firm. We never lose contact with reality, and we do not underestimate our social responsibilities. As we have over one hundred active clients on our list, it may well be that the soundness of our designs affects the lives of millions.

I spend a great deal of my time in plants and I enjoy it. There is something about a large up-to-date American plant in action that deserves the descriptive attention of some great American writer. It is one of the most exciting sights in the world and one that America alone can offer. A remarkable thing, and one that impresses me still after thirty years, is the natural elegance of the American workman. I don't mean cab drivers or longshoremen, but factory workmen, in their well-cut overalls, gauntlets, and peaked caps; they look like magnificent ambassadors of a great industrial nation, and they are "representatives" in the grand manner.

I have seen automatic-machinery operators, assembly-line or spray-booth operators that would make movie stars look like tired headwaiters. Sometimes welders, their helmets raised over their foreheads, in light overalls and black gauntlets, are a valid expression of manly elegance. It is a familiar sight around plants at shift-time to see workmen leaving for their homes in smart streamlined jobs, tanned, neat-looking, and athletic. They have poise and dignity, they have a cheerful look on their faces, they are gentlemen. These fellows, young or white-haired, represent the working aristocracy of the world.

I often daydream about such a fellow riding in some foreign city in his 1950 beige landcruiser, a cigarette in his mouth, radio full-on playing the latest dance hit, and well dressed. He stops near the curb.

"Who are you?" asks a passer-by.

"My name is Smith. I am a drill operator in Plant 6 of the General Rotors Company. I am on my way home to install the new dishwashing machine I gave my wife for her birthday. Must get along now or we'll miss the six-thirty television show. S'long, brother."

It would be great propaganda for the American standard of living, but they wouldn't believe it. They would think it was "a plant" and all phony.

CHAPTER XIII

SUDDEN VARIATIONS IN SIZE IN ORDER TO BREAK THE MONOTONY OF A REPETITIVE DESIGN HAS ADVANTAGES. IT CAN BE ACHIEVED EITHER THROUGH SHEER MAGNIFICATION

OF THE VERY SAME SYMBOL, AS ILLUSTRATED IN THIS TEXT, OR THROUGH A SHIFT TO

A COMPLETELY DIFFERENT SYMBOL AS EXPRESSED NOW IN THESE PRINTED WORDS. IN COMBINATION WITH COLOR AND TEXTURE, IT CAN BE USED EFFECTIVELY.

IT SHOULD BE UNDERSTOOD THAT THESE EXAMPLES ARE MERELY AN APPLICATION TO THE TYPOGRAPHICAL FIELD OF A TECHNIQUE THAT IS ORDINARILY APPLIED TO PRODUCT DESIGN.

Chapter 13

MICHAEL AND VENISE

When Germany attacked Poland, I felt that the world I had known would never again be the same. Then France was invaded and the situation became worse from day to day. While in Dayton, at the Frigidaire plant, my engineering friends were as sad as I was at the dreadful news. When I left them, a boy selling newspapers at the plant door was shouting, "*Paris falls!*" I bought a paper; it was true; my Paris had been taken by the enemy. To Jean, my wife, who loved France, where she had spent her youth, and to me, who was born there, it was just agony.

In 1940, when London was being reduced to ashes by the Luftwaffe, I received a cablegram from my English friends the Ashley Havindens. They were asking me whether or not I could keep their two young children for the duration. Jean and I cabled our acceptance immediately. A fortnight later, I left Jean on board our boat, the *Loraymo*, on which we were cruising near Boston, and went to wait for them in New York. Their ship landed in Baltimore and they flew in to La Guardia Field. With a certain emotion, I had bought a doll for Venise, age ten, and some candies for Michael, twelve. I had prepared a rather sentimental little welcome for two such young children away from home; all about hands across the sea, my home is your home, victory, and all that sort of thing. When I saw the kids I rushed, ready to give them a big, hearty American welcome. Michael did not give me the chance: he held out his hand, said hello, and asked me who was pitching for the Giants. Apparently he had learned plenty aboard the ship during the crossing. Then I kissed little Venise. I asked her, "Venise, how was the trip?"

"Dull."

I offered the doll.

"Do you like dolls?"

"Not very. Do you?"

I turned to Michael.

"Well, Michael, are you happy to be here?"

"Yes, quite."

Then: "Raymond?"

"Yes, Michael."

"Can I ask a question?" (All that with a terrific British accent.)

"Why, of course."

"What are Mr. Willkie's chances?"

I tried to answer to the best of my ability. Then silence again. This time it was Venise, with that same very, very British tone.

"Raymond?"

"Yes, Venise."

"What are you going to do in America about the Negro problem in the South?"

In self-defense I bought them both double ice-cream cones so they would keep quiet for a while until plane time.

Silence while we waited for the Boston flight. The warm little spiel of welcome I had prepared was put on ice, and it is still there.

Venise and Michael turned out to be adorable children. They became self-Americanized, without pressure or propaganda. High school did it, not some fancy institution. At the Port Washington (L. I.) High School, Venise was chosen by the other kids to carry the flag in the Fourth of July parade. She cried with pride and joy. Jean and I nearly did too. Michael was extremely popular, and he was elected head of his class by the other pupils. He loved baseball and he reached the point of exuberance (pretty bad for a young British gentleman) on the day my friend Bob Considine took him to see a ball game and had him meet Joe di Maggio. Michael and Venise stayed with us more than three and a half years, then went back to England in 1944.

The life of Jean and myself at home in Sands Point was never the same. One thing led to another, and eventually to a divorce that has left us genuine friends, with a great deal of mutual respect.

Three years ago I was in London on business. I went to see Michael at Harrow. He had become a handsome young gentleman, six feet tall, and president of the students' organization (or whatever the exact title is at Harrow). He was also chosen as one of the two boys, out of the whole student body, to take part in the annual debate. This major event takes place in the same auditorium, on the same podium, where Disraeli and Churchill debated in their time. Michael, a really brilliant young man, won a fellowship at Oxford, where he is at present.

Venise has become a first-rate water colorist. Her work was among the very few pieces that were selected by the government for a world-wide traveling exhibition of British modern art. Both have a real fondness for America, which shows their intelligent appreciation—because, as someone else has said, "It is proof of mediocrity to admire in a mediocre way." I am very proud of the kids; there is nothing halfhearted about their admiration for our country.

Chapter 14

VIOLA ERICKSON

Some observers have attributed my relative success in life to three major factors:

a. An unbounded imagination.
b. Luck.
c. Determination.

Whether or not this is true businesswise, it certainly applies to the most important aspects of my personal life. In my wildest daydreaming I formed a conception of the perfect woman. It was an utter sublimation of everything that meant gentleness, beauty, charm, vivaciousness, intelligence, chic, etc., etc., and a thousand more et ceteras.

Luck, unbelievable luck, brought that dream to me. My determination did the rest. I married my adorable young wife Viola on December 22, 1948.

Until I was about fifteen years old, mine had been the life of a frustrated young perfectionist. Erringly, but intensely, I was groping for perfection—perfection of color, of rhythm, of warmth. I loved everything gay, luminous, and fragrant, everything alive and vibrant. I felt it existed but I did not know where—and in what form, or feel, or sound. I was too young to be analytical about it, and I merely hoped that someday perfection would reveal itself in its radiating beauty in a simple, unexpected, final form. In a juvenile way, I had looked for it in more or less obvious places, whether at the Louvre, in the park of a château in Touraine, in meadows and forests, at concerts, or near the sea. I was moved by the splendor of the world that God had made for us, yet I felt vaguely unsatisfied.

Many times I felt that I was near my perfectionist's goal. Once it was in a tall forest of Île de France. It was after a shower; the sun was now shining, low at the horizon. The lazy rustle of raindrops falling from leaf to leaf, the scent of mushrooms in the cool October evening, a hunting horn far away, all blended exquisitely in subtle half-tones. It was perfection—short of two things: warmth and life.

Many more times I felt the same closeness to my ethereal target. A certain spring evening in Normandy, I stayed overnight in a thatched-roofed village near the place where Marcel Proust spent his adolescence. Alone, as always, I was drifting leisurely in the somnolence of the village. The small church at sundown was a gem of Gothic preflamboyant purity. I entered the penumbra and looked for a pale of darkness. I leaned against the cool limestone and admired in silence the stained-glass windows ablaze with the evening sunlight. I remained there a long while, relaxed and happy. Then a gaunt, white-haired gentleman emerged from

VIOLA LOEWY. *Jean Du Val Photo.*

the dark transept and sat at the organ. He played. Softly, for God and for us, and we three were alone. He played beautifully. The whole thing expressed absolute aesthetic elegance. I thought I had found my goal, but not quite, I felt. It was spiritual and aesthetic perfection, assuredly. But *my* goal was something else. What could it be?

Shortly after my fleeting but lovely encounter with the lady in the train, while returning from school, something had happened deep within me; my conception of the goal gradually became far less ethereal. It wasn't long before it showed definite traces of a waistline, silken hair, and cherry lips.

So that's what it was! Well, well, well!

I double-checked with my earlier specifications of the goal—color, rhythm, feel, fragrance, luminosity, vibrance—it was all there, all right. And it was warm too. I had found perfection at last, and it was the Woman.

From that day on life had a new meaning.

My admiration for the human female has grown daily more absolute, and those extraordinary specimens are so unpredictably adorable that I have found life away from one utterly miserable. The presence of a well-formed and well-eyelashed young female, whether gowned by Dior or wrapped in a piece of burlap (or even just plain), is enough to make me happy. My inspiration is she. My fun is she. My desire to live is she. She conveniently happens to be my wife—Viola Erickson Loewy.

The place occupied by the female in my existence having been established, it may be fair to admit that, in this respect, life has treated me well. I am one of those fortunate men whose experiences with women have been consistently happy ones. Even when fate disrupted my first marriage, it left both of us unscarred and good friends. As a result, my feelings for the specie are a mixture of adoration plus respect. I feel sorry for the man who possesses either one or the other, but not both. He will never know the meaning of absolute love.

Careerwise, this understanding and appreciation of women has been helpful. It has been said that the products we design reflect this feminine comprehension. Without ever getting frail or too delicate, our designs nevertheless retain a certain feeling of slenderness and grace that appeals to women. It fortunately happens that most men like these attributes too. Even in a steam shovel or a ten-ton truck, organized design gives the whole thing a quality that appeals to their innate leanings for orderly power and thoroughbred strength.

Jean and I remained married for fifteen years. Together we went through the whole cycle of our business development, from a one-man show to what Raymond Loewy Associates is today. Jean is a brilliant woman, erudite and with sure taste. Together we were happy, and she was a charming companion; we had good fun together during all those years. Her intelligence, her charm, and her wit were great assets in my career. No important decision was taken without consultation with Jean. As mentioned before, war and fate (in another form) parted us.

We have been divorced for nearly five years, but she has remained an active partner of the firm. Important issues affecting policies are discussed by telephone, and she makes periodic trips to the States from her home in Paris.

Jean is now married to Jacques Bienfait, a Frenchman, father of three small children. Jacques is witty and quite handsome. I am happy to see her married to such a charming man. When in Europe, Viola and I enjoy seeing the Bienfaits and their kids, who are a lot of fun.

It is my good fortune that Viola is also an extraordinary person. She is not alone a most intelligent young lady, but also a sincere and a fair one. As a result, she understands what Jean Bienfait meant in my early life and what we went through together. She understands that these feelings cannot be obliterated, as they are based on mutual respect, on a long teamship in a hard world. Viola has a real affection for Jean. Jean, in turn, thinks Viola is thoroughly enchanting and the perfect wife for me. So we are all good friends, we adore the Bienfait children, and we all live our parallel lives—they in Europe, we in America—without rancor, regrets, or bitterness.

By 1939, I had achieved some relative success that brought me unexpected but pleasant rewards. The British Royal Society of Arts made me a Fellow (FRSA) and the title of Royal Designer to Industry (RDI). My friend the brilliant Georges Doriot, of Harvard's faculty, invited me to lecture at the university. In 1941 the United States Department of Justice selected me to deliver a coast-to-coast broadcast as the nation's naturalized-citizen speaker for I Am an American Day. I was introduced as follows: "The Department of Justice of the United States presents a broadcast for all Americans. Today it has as its guest speaker a man who has left his mark on our contemporary American way of life, Mr. Raymond Loewy, the industrial designer."

I had tried to be a good American, to contribute my humble share to its greatness, and I had succeeded. What a marvelous feeling! I felt proud and happy.

In quick succession, I became a member of the Advisory Board of New York's Board of Education, a fellow of the Society of Industrial Design, and later its President. On the day the American troops entered liberated Paris, the Department of State paid me the great compliment of giving me the opportunity to speak on the radio to the French people, in French, as an American citizen. Finally, I had the thrill for a naturalized subject of being received in private audience by the President of the United States at the White House.

On that spring day of 1947 I remembered with emotion my entering America in New York, twenty-eight years before, and a certain visit to the top of a skyscraper.

It was such a long, long way looking back at these early days, but it had rushed by so fast. I was now an old New Yorker; a fellow who had known the Knickerbocker Bar, heard Gallagher and Shean, whistled "Dardanella," known Tex Rickard and Harry Greb. I understood America better and liked it more every day. I was still impressed by the contrast with the old world. In what way?

Take charity, for instance. The average American would give his heart away provided nobody looked, but he is reluctant to be caught red-handed in the act of being charitable. It is a form of decency. Certainly there is nothing like that in most European countries.

Then there is the American theatre, which happens to be a respectable field of operations for respectable people. Whether they are stars or plain show people, they have dignity. Morally and physically, the theatre is clean and wholesome. I would hate to start comparing it with the theatre of Continental Europe, which is often sordid, dirty, and vice-ridden.

The contrast would be even more striking in the case of the press. For those who criticize American journalism for occasional inaccuracies or spectacular displays, I recommend a study of European dailies. They would quickly appreciate the splendid job done by such national institutions as the New York *Times*, the *Herald Tribune*, the Washington *Post*, the Denver *Post*, and the thousands of other papers that operate on a basis of common decency, fair play, and business integrity. It is my good luck to know a great many journalists, and I find them a brilliant, witty bunch of swell fellows, wide awake and alert. For speed and initiative, they are hard to beat.

This also applies to the magazine field—perhaps the most sophisticated professional group that I have had the privilege to meet. There we find a fascinating combination of talent, humor, and imagination. Some business luncheons I have had with staff writers of famous news magazines remain exciting experiences of my business life. The variety of their knowledge, the keen insight of these men in an infinity of fields, even highly technical ones, are simply amazing. One must know them to understand the success of the publications. Driven by envy and devoid of scruples, countless "publishers" try to copy them all over the world and they fail with impressive regularity.

The headquarters of American fashion magazines are very exciting places indeed. The atmosphere is a melange of neo-nonchalance, sophomoric wisecracking, and Chanel 5. Offices and corridors are teeming with chic young things, all permanently hatted, most of them a bit "diverged." They manage to do a beautiful job in constant oscillation between their latest nervous breakdown, their next passionate crush, and the fall-fashion issue. Without their trade, Park Avenue psychoanalysts might as well hock their couches.

One may, occasionally, grow a bit impatient at the didactic finality of their captions, but all in all they are doing, aesthetically speaking, as much good in their field as the *New Yorker* does for humor and the preservation of American finesse.

As far as the art directors are concerned, I consider them the leaders of their profession, whether here, in Europe, or anywhere else. Men like Lieberman, Eude, Brodovitch, Eltonhead, and Armitage are all artists of exquisite taste. May I pay my respects to my art director friends who are doing so much to raise the aesthetic standards of America.

Some traits of our land, however, are a bit irritating; such, for instance, as the attitude of the policeman giving you a summons. The rather friendly and chummy attitude of the motorcycle cop who hands me a ticket while calling me "Bud" doesn't digest well. I'd just as soon be given complete hell by a Parisian *agent de ville* who finally tears the ticket and tells me to drive on and watch out or else. Or the apathetic indifference of the public toward the great American Laundry System, whose destroying potential is simply staggering. The brains of this industry ought to be removed to the Pentagon Building and be placed in charge of the chemical warfare division, or anything pertaining to complete and final destruction.

Equally surprising is the patience with which the public tolerates the defacement of most American towns and villages by ugly telephone or power line posts. I have seen dozens of communities which were aesthetically raped in public by these monstrous contraptions. Frank Lloyd Wright and I agree on a few things and disagree on many. But when we fight, our common hatred for defacing telephone poles is such that we can always make peace by bringing up the horrid subject as common ground for armistice.

As to the well-known argument about whether or not American movies have a nefarious effect on the population, I have made myself quite unpopular with countless people by taking a firm stand in favor of the movies. Thanks to the motion pictures, most Americans down to the smallest farming belt or coal town have learned about table manners, correct clothing, good music, etc., etc. They have made it possible for such firms as Sears Roebuck and Montgomery Ward to sell, even in the tiniest farm or suburban cottage, three-ninety-five dresses well cut and designed by the foremost fashion designers. What a contrast with the miserable clothing of the European peasant or worker! It also helps to sell millions of home appliances and labor-saving devices which Mary Smith or Bill Jones saw in a kitchen, garage, or nursery scene of some movie. All this is excellent for the maintenance of the standard of living that Europe envies and admires. As far as spiritual values are concerned, I can't see that the films make any difference one way or another.

In my opinion, the American movie is a great factor in the economics of the nation, and a helpful one. Abroad, it is a major means of effective propaganda for the American standard of living. Whether or not they get an intellectual lift from the plot, the foreign customers certainly enjoy seeing Dagwood's new Chevvy, Blondie's Frigidaire, and baby's plastic diaper.

All this is obviously no excuse for inept plots, ham acting, and "arty" direction. But the situation is getting better; the industry is slowly freeing itself from the sad influence of Cecil B. de Mille and the like with all the vulgarities it implies. The most objectionable thing about movies is not on the screen; it happens in the audience. Anyone who has been sandwiched for three hours between two kids munching peanuts with their mouths half open will understand what I mean. The warm stench of gooey peanuts is deadly.

Another interesting contrast with Europe is the general disinclination of the American to repair anything out of order that could just as easily be thrown away. This wasteful attitude is only wasteful in appearance, for it increases turnover immensely and therefore increases production and sales, creates business, and adds employment. Should the public adopt the European attitude (created by inborn inclination and economics) of hanging on to a given product until it completely falls apart, our economy would be much different. One can easily see what would happen in the automobile business if American consumers wanted to retain their cars for ten or fifteen years, as many Europeans do. Production would take a nose dive and the whole economy would suffer. My friend Paul Hoffman once said that this difference of attitude explains to a great extent the contrast between American prosperity and European penury.

Naturally, this nation-wide tendency to discard anything in need of repair can be carried to extremes, as in the case of married life, where many situations that would successfully respond to the patching-up treatment are just thrown in the junkyard of the divorce court.

One of the pleasant surprises that struck me when I arrived from Europe was the fact that one so rarely sees corpses in the street. Anyone who has lived abroad knows how often one encounters funeral processions slowly ambulating through the streets of a large city. It is not infrequent to see two, three, or more in one day. It does not do much good to the soul and it is bad for traffic.

Here, one seldom sees anything like it except on special occasions, such as the obsequies of an Italian gangster or a Tammany Hall big shot; death has been streamlined. By contrast, I have seen many a sleek hearse-limousine driven, cold load and all, in such a masterly way that it managed to avoid all the red lights, beat the green ones, and make the Fifty-seventh Street–Woodlawn run in thirty-seven minutes flat. As a lover of speed, it is my faint hope—should I end my days in America—that I will take that last ride in a hot, souped-up job. I can even conceive being arrested by a motorcycle cop who, trite to the end and with poor-taste finality, will ask the driver, "Where the hell do you think you are going?" and hand him (for doing sixty-five on Queensborough Bridge) his latest ticket and my last summons.

America's informal approach to the business of the kick-off-permanent sometimes takes aspects surprising to the visitor. Such, for instance, is the case of a blatant third-rate night club on a main suburban artery that met with some bad luck and folded up, possibly on account of its poor location—right across from a cemetery. The owner had spent plenty of G's on a sprawling black-mirrored façade decorated with several large circular windows framed in gilt. It was loud and real gaudy. A jumbo-size neon sign shrieked CLUB TAMPA, until a placard said: TO LET.

A certain casket manufacturer rented the place and put a sample "slumber-case" in each porthole, barely hidden behind an opened Venetian blind. All he had to do was to change the sign from Club Tampa to Club Casket Company.

This is all very nice, but there are limits to bad taste. The limit is the famous "Slumber Gardens" of Los Angeles which Evelyn Waugh, the English novelist, took as a motif for his lethal satire *The Loved One*. A great part of his material, unfortunately, is authentic and it makes us look ridiculous to the rest of the world. The mixture of death and humor is one thing; that of death and vulgar sensualism is another.

The British themselves are not without peculiarities on the subject. On my last business trip to England, while two of my designers and I were driving from London to Birmingham, we saw a simple countryside cottage called Chisham Arms. It was divided into two parts, the façade being painted white on the left and pale blue on the right. Over the door on one side was a sign:

RESTAURANT GRILL; over the other: CREMATORIUM. All very nice and chummy . . . I can imagine one of those perpetually rushed American tourists in search of a quick hamburger walking inadvertently through the wrong door and barking, "Make mine rare with plenty of mustard!"

Chapter 15

PREPARATIONS FOR POSTWAR

This is war, total war. My friend Donald Nelson, of Sears Roebuck Coldspot days, is now trying to make a military nation out of peaceful America. As head of the War Production Board, his task is to transform American industry into a gigantic source of war material. I could not imagine a better man or a more courageous one for the staggering task. The fact that he was in charge gave a boost to my morale and renewed the confidence I needed so badly; France was invaded, London in flames, most of Europe occupied, bled to death and terrified. Then Pearl Harbor. We were licked on the seas, licked in Bataan, in the air, and in Guadalcanal. Our mobilized men had no weapons to train with and they used broomsticks as rifles or empty crates for artillery. Donald Nelson, confronted with a superman task, kept his calm and his lucidity—and did the job.

In March, 1950, he came to visit us in Palm Springs with the charming Mrs. Nelson. We drank a cup of champagne to the memory of those heroic days, and to Donald, a very tired man. I am happy to have an occasion to express, as every American should, my gratitude for what he did for the nation.

The war period was a test of the industrial designer's ability and resourcefulness under handicaps of all sorts. Civilian manufacturing had to be kept running in spite of shortages of all sorts: shortage of skilled labor, materials, and even machines. It was necessary to keep on making the essentials of civilian life. After all, 80 per cent of the population never left home and life had to carry on. Besides, it was imperative that working people not required in war plants should continue being employed, fed, and even entertained in order to keep up the morale of the nation.

Designing products without steel, copper, zinc, tin, or lead, giving them appearance without paints, lacquers, or dyes, was no cinch. Even plastics were often on the restricted list, and what was left was not much.

By using plenty of imagination we were often able to keep a plant running. Take the lowly lipstick container, for instance. This apparently unimportant item was in reality a factor in maintaining a high level of morale. American women without lipstick would become highly vulnerable to gloom and dejection. Men would follow. We went to work, and after days and nights of experimentation we were able to develop not only a lipstick container but a swivel type that required no metal of any kind. It worked well and the manufacturer sold hundreds of millions during the hostilities.

There were many other examples of our role in keeping things running during the emergency. This we did while working on the development of such problems as a glider containing a complete field hospital that could be flown, landed, and erected in a matter of hours, in the battlefront's critical areas. Also for the Medical Air Corps comprehensive emergency first-aid kits that could be parachuted without breakage. For the Ordnance Department, fast-loading equipment for coast artillery guns. For the Air Corps, an idea for a controllable rocket-operated flying target—this as a device for the training of fighter pilots. Also swamp shoes, tents, and a dozen other items for the Quartermaster Corps. We camouflaged enormous plants, such as Glenn Martin in Baltimore.

In 1942 it became clear that industry should get prepared for the postwar period in order to be able to absorb the demobilized G.I.'s as fast as they could be released; also in order to reduce or eliminate a potentially disastrous production gap between war production and postwar manufacturing. This had to be done without interfering in the slightest with war production. In fact, it was often impossible to even reach the engineers, and we did our best to go ahead on our own. For instance, when the war ended, we had succeeded in developing a complete line of entirely new automobile bodies for Studebaker, ready for the tool and die makers. This was done by my own designers, men either beyond draft age, heads of large families, or physically handicapped. It was done without using any critical materials and naturally without any guidance as to what postwar automobile styling was likely to be. We used our own judgment, and we were lucky enough to guess right. However, our luck was helped to some extent by a certain amount of logic which we applied to the problem. Simple, sound factors such as better visibility, lighter weight, and fast, slender appearance. The success of the car, with improvement in the standing and financial structure of the company, seems to substantiate the accuracy of the reasoning. Studebaker came out with a new postwar automobile more than a year before the competition. R. L. A. managed to place many of its clients in a similar advantageous position.

At the end of the war, membership in the industrial design profession had reached nearly two thousand. Unfortunately, a great many of these men were not qualified to practice such a highly technical profession. Dreyfuss, Teague, and I got together in order to discuss the situation thus created and endeavor to take steps to correct it.

It became apparent that an organization was needed, some sort of Society, whose membership would require the meeting of certain qualifications and professional standards. We invited ten other leading designers to join us, and in 1944 the Society of Industrial Designers became a reality.

Our first task was to establish a basis for admission to membership. Then we took in members who fulfilled the requirements. The Society soon began to grow and it now has more than a hundred members, all regular industrial designers. It was my privilege to become a fellow in 1944, and its President in 1946.

As head of the organization, I felt that my first duty was to establish a code of ethics. A committee drafted it and it was completed and adopted during that year. It has done a great deal for our Society.

Somewhere along the line the S. I. D. changed the name of its dictum on fair play from the "Code of Ethics" to the "Code of Practice," which, it appears, is the accepted circumlocution. This now states that a designer shall be honest, loyal to his client, and not work for two clients in direct competition.

Further, the designer shall not injure the business of a colleague with malice aforethought, shall not try to sneak said colleague's clients out of the side door, and shall not steal his designers.

As if all this would not burst the seams of mere man's moral frame, a designer shall not make prophecies, shall be accurate in claiming design credits, shall not advertise, shall not enter contests unless approved by the Society (in practice, therefore, it is unlikely that an industrial designer will ever be an Olympic swimming champion), shall not set up one-man exhibits of his designs without permission of the Society.

All this appears to be generally confining. The truth of the matter is that—with minor exceptions—this has been the unspoken code since the beginning of industrial design time. So stated, it might seem that designers could very well spend the majority of their productive lives standing in the corner for being bad boys. Not at all. Industrial designers are exceptionally ethical in their practice. They behave well.

Membership in the S. I. D. is just a little more exclusive than in the Explorers of Eastern Somaliland. One must be a practitioner or a professor of industrial design possessed of ability, integrity, and character. In the case of designers, they must have carried major design responsibility in the designing of three or more different products which ended up on a sales counter somewhere, must have worked for three different manufacturers, and must be sponsored by at least two Society members.

Besides headquarters in New York, the S. I. D. now has chapters in Chicago, Detroit, and Los Angeles.

PART THREE

Let us open the last part of
this book with a silent prayer:

MAY AMERICANS, MY COUNTRYMEN, SEE THE LIGHT AND NEVER AGAIN PUT MAYONNAISE OVER FRESH PEARS.

AMEN.

Chapter 16

THE NATIONAL WIDGET COMPANY

With the end of hostilities, skilled labor, machines, and materials became available again, and we resumed our design work without handicaps or restrictions. The war years had taught us the value of inventiveness and economy of materials. Industrial design became even more efficient and cost-conscious: it had graduated from an experiment by a few enthusiastic men into an accepted and respected profession.

Today no manufacturer, from General Motors to the Little Lulu Novelty Company, would think of putting a product on the market without benefit of a designer. It does not matter whether he is an employee on the firm's own staff or a free-lance consultant. This in the space of twenty years. I take it as a complete vindication of my early theory that, eventually, correct visual presentation would become an integral part of merchandising in practically every field.

In the postwar years, the attitude of the prospective client when confronted by an industrial designer in search of a contract is quite different from what it used to be in the early thirties. He has heard and read about industrial design; the profession is not a mystery any more. But he doesn't realize how it can help his own company. He makes nice Widgets, they sell all right, and he doesn't believe he really needs any help from outside. However, several of his top distributors, the advertising agency, and a couple of directors of the company have mentioned to him a certain industrial designer who seems to know what the public wants.

"This fellow has designed the new 1949 Blidgets line, and sales went up 200 per cent," they say. "Why not write him a note and ask him to drop in next time he's in St. Louis? Wish you'd try, Jack."

Jack Smith thinks it over for a while and finally makes up his mind. He writes a letter to the designer suggesting a meeting for a discussion of his company's design problems. In cases where the firm has never worked with industrial designers, the interviews usually run somewhat as follows:

Jack Smith: "Mr. Loewy, I know of your reputation and of your achievements in your profession. In fact, I have asked you to come and see us at the request of several members of our organization. We are the largest Widget manufacturers in America. We know that you have designed successful trains and ships, airplanes and business machines, and many other things; but how do you know that you can help us to solve our problems? You may not realize it, but Widgets are very special things, in a class by themselves. Some of our engineers have spent a lifetime designing better and cheaper Widgets. What makes you think that you could improve on what they have been doing?"

R. L.: "To be exact, Mr. Smith, we have designed a successful line of Blidgets for the Thompson people, and they are quite similar to Widgets in many respects. However, in order to try to convince you, let us assume that we never did design anything remotely Widget-like.

"My associates and I are familiar with the National Widget Company. We have studied its records for the past few years. We know the reputation of your firm for business integrity and the quality of its products. From a merchandising viewpoint, however, we believe it could be improved. Your present models seem to lack certain physical characteristics that would make them stand out among the competition. For one thing, they might reproduce better in your newspaper advertising. The present models are rather weak in appearance and they lack sparkle and highlights. We feel that a competent outside organization with design imagination, working in close co-operation with your engineers, might develop a fresh and unusual answer to your problem."

Mr. Smith: "How would you propose to tackle such a problem? What would be your suggestions in a case such as ours?"

R. L.: "First, we should become thoroughly acquainted with your plant and equipment, so that whatever design we submit to you is within your plant's manufacturing limitations. Parallel with this, you might be interested to see some advanced designs that would possibly involve new tooling or equipment. But this we can discuss later.

"Second, we will make a study of the Widget market and of your competitors. Should there be some outstandingly successful Widget manufactured by another company, we shall try to isolate the reason, or reasons, for its success (whether it is price, appearance, performance, quality, etc.). In this, we would expect the co-operation of your sales department. Then we can talk to you intelligently about your design problems and suggest a fee."

Mr. Smith: "I'll tell you frankly, Mr. Loewy, that it all seems quite logical, but it also goes a great deal further than what I had in mind. I had more or less assumed that you would simply make some sketches giving me your idea of what our product ought to look like. This, Mr. Loewy, was what I . . ."

R. L.: "Well, Mr. Smith, I regret to say that I don't believe we could be of assistance to you in that direction—or any other reputable industrial designer, for that matter. We would be most hesitant about submitting to you any design that hasn't been developed in full knowledge of the factors involved. I don't think you would do it either, Mr. Smith."

Mr. Smith: "You are reasonably sure that you could find out enough about our own particular field in a short time to be able to suggest improvements to our products, then?"

R. L.: "Certainly, provided we get the co-operation of your various departments. You see, Mr. Smith, Widgets, after all, are not so very different from many other items which we have designed in the past. These items may not actually be Widgets, but their manufacturing, selling, and servicing are very similar. Even pricewise, Widgets are very close to Blidgets. I can assure you, Mr. Smith, that with your engineering department's co-operation, we will be able to increase the acceptance of your product, and possibly reduce its cost."

Mr. Smith: "To do so, then, would mean taking you and your associates into the family, so to speak. We would have to divulge to your firm many confidential matters and business methods."

R. L.: "Yes, completely. Please remember that our professional reputation is based greatly on our past record for discretion and ethics. Some of the largest corporations in the land have taken us into their confidence for fifteen years or more. You can rest assured that any development taking place during our association will be protected by us and kept in absolute confidence."

Mr. Smith: "I am quite sure of it. But tell me this. In such an association as you refer to, aren't you afraid that we may encounter some clash of personalities? We have to consider the human angle, you know. Some of our engineers have been with us for more than twenty-five years; they have grown with the firm. They have a certain pride in their work, and, if you ask me, they know their stuff. Don't you think they might be a little sensitive about outsiders' being injected into the picture? I am wondering whether or not we might not have some sort of conflict when you people come into our organization."

R. L.: "I don't think so. We have met such situations in the past, and they never became a major issue. If you would consider just leaving it to us, you'll see that there won't be such problems."

Mr. Smith: "Remember that our Widget business is different. It's not like any other business, you know."

R. L.: "You can't be more difficult than other people we have successfully worked with before. And, to me, you don't look like unreasonable people that will be difficult to get along with."

Mr. Smith: "Well, I guess not, Mr. Loewy."

R. L.: "I know I am right. Besides, look here, Mr. Smith, we are *not* interfering with the work of your engineering department. These men are bound to find this out very quickly. We are more than willing to give them all credit for whatever they contribute during our co-operation, and they will soon realize that we are not after their job. Why should we be? We are doing pretty well on our own, you know. And when the result of our joint efforts is placed on the market and becomes a sales sensation, we shall make it a point to give your men all due credit. Whatever press release is distributed by our public relations department, after *your* okay, will mention the fact that the new Widget was designed by Raymond Loewy Associates *in collaboration* with the Engineering Department of the Company. No, Mr. Smith, you have no cause for apprehension on that subject. In fact, you may find out that members of your own staff are the very ones that will ask you to call on us when the next problem comes up."

Mr. Smith: "Mr. Loewy, you have clarified many points and I am inclined to go along with your ideas. However, as President of this Company, and being aware of my responsibilities toward the stockholders, I am wondering as to what are the guarantees of success. After all, all this is going to be costly, and I must have some reasonable assurance that we will get adequate results for our investment. It isn't as if you were merely to prepare some sketches for us. Now we are talking about a rather ambitious program and we must get an adequate return for our money."

R. L.: "Well, I don't see how you possibly could ask, say, a surgeon, even a top-flight one, for a guarantee of a successful surgical operation. All he could tell you is that the chances of success are great, moderate, or practically certain. In no major case could he honestly guarantee success. As far as I am concerned, all I can say to you is that our record of successful designing speaks for itself, that in this case the 'operation' is not an unusual or complex one, and that the patient looks quite sturdy. The chances of success are great, and there will be plenty of new little Widgets in the world. Should it make you feel reassured,

I could give you a list of nearly one hundred corporations by which we are retained. Among these you are bound to have friends. You might wish to call some of them to ask, 'How did you get along with the Loewy boys? And what about results? Did you get your money's worth?' "

Mr. Smith: "Then, you wouldn't mind having us contact some of your clients?"

R. L.: "Oh, not at all. In fact, I would welcome it. Here is a list of our clients as of this date. Please keep it."

Mr. Smith: "I must say that you sound rather convincing, and we might like to try for a month or so, and see how we make out."

R. L.: "Well, that is difficult. I don't see how we could possibly function correctly on such a short-term basis. You see, if we start with you, it means that we cannot work for competitors of yours. That is the ethics of the profession. We could hardly consider an association shorter than for one year."

Mr. Smith: "If you were to acquaint yourselves with our operations I suppose it would take some time before you would get the whole picture. Then it would represent a substantial investment and we would want to get the most out of it. I can see that we should make our arrangement valid for a reasonable length of time so that we can find out how it is going to work out."

R. L.: "There is another possibility that might make it easier for you to accept, Mr. Smith. If you hesitate to commit the company for a year's fee, which you say might be difficult to justify to your board of directors, why not make another kind of deal? Suppose you retain us for a period of twelve months at a nominal fee of, say, a few thousand dollars, and then give us an extra compensation, in the form of royalties, based upon the success of our designs. Your fixed investment would then be reduced to a minimum. As far as we are concerned, we would welcome such a deal. It has been very successful in the past."

Mr. Smith: "You seem to have a lot of confidence in your ability to design more salable Widgets. But I am not ready to go along on such a basis at this time. Perhaps we should consider instead a reasonable fixed-fee arrangement."

R. L.: "Glad to."

Mr. Smith: "Well, you have given me a new outlook on the whole business. Perhaps, after all, industrial design can be of value to us. Why don't you write me a proposition for a period of one year, covering the design of our complete line of Widgets, including the packaging?"

R. L.: "Shouldn't we also include point-of-sale displays and other merchandising requisites?"

Mr. Smith: "Yes. Also, you might have some ideas for a new trademark that could be used on our delivery trucks and our stationery. It hasn't been changed for twenty-five years and I guess it could be improved."

R. L.: "Okay. You will get it within a couple of days."

Mr. Smith: "Right. I will take it up with my associates and we will let you know shortly. If we are to go ahead, we don't have any time to lose as we must have our new '52 models ready by November at the latest."

R. L.: "We can get started any time you say."

Mr. Smith: "That's fine. Here; take that Widget along with you and try it. I think you'll like it. Thanks for coming to see us."

This is rather typical of what happens in the field. Once a contract is signed, it doesn't take the designer long to convince everyone that he knows his business and that he isn't after anyone's job or position. In fact, it acts as a morale booster, and often the engineering, research, production, and sales staffs get sparked up and alive again. The results are usually beneficial to everyone concerned.

Our design fees vary according to the difficulty of the problem. We take on design jobs for as little as $500 or as much as $300,000. If the unit to be redesigned is a big thing—for instance, a tractor—there are so many obvious things one can do to make it better-looking that a relatively low fee is in order. But if we were to redesign a sewing needle, I'd charge $100,000. After all, it isn't simple to improve a needle. It's like the perfect functional shape of an egg.

The majority of industrial designers operate in the realm of products design: that is, household appliances, hardware and utensils, electrical and electronic products, machinery and equipment, etc. Some of them extend their services to packaging and displays. A few are also engaged in the planning of small stores and shops. A very limited number are consultants in the field of transportation, designing interiors of commercial airliners, ships, trains, or busses. In addition to the above, the larger organizations include a section of department store design. This division, which we call Specialized Architecture, is fast becoming an important one. We shall describe it later in more detail.

Beyond these activities normally carried out in my company, we are constantly looking for more terrains of action. The reason for this is that the majority of America's important business, industrial, or research organizations require in their operations something that we can supply: imagination and a fresh viewpoint. This, coupled with a sense of realities and a reputation for being discreet in confidential matters affecting our clients' most secret projects.

The top executives of such corporations must naturally be gifted with plenty of imagination themselves in order to discharge their duties successfully. They also have experience, in fact, enough of it to realize that imagination is at a premium and that there is always a need for more in a well-managed company. As a result, we are called upon to collaborate in the solution of problems that, at first glance, would seem far remote from industrial design in the sense of visual engineering. For instance, there is the case of a leading smoking pipe manufacturer who asked us to have a look at their manufacturing plant: not in order to redesign the product, but simply to obtain an outside reaction to their operations from men with a reputation for good taste, business sense, and practicality. Or the case of several magazine publishers desirous to have consultation about their periodicals' layout, type-setting, and physical appearance in general. Recently, a well-known manufacturer of girdles and brassières asked us to become part of his organization's research unit, our role being to keep an open and alert mind about new types of feminine equipment. In this particular case, the problem could be reduced to a few constants: hips should be made smaller without transferring the excessive bulk to parts of the topography where it would be even more unwanted. This problem of transportation, distribution, and camouflage is engineering.

Brassières should do an efficient job of uplifting without betraying the presence of girders, supporting members, or other hoisting implements. This represents a problem of cantilever construction familiar to both bridge designer and stress analyst.

The female in general seems to be confronted with perennial problems connected with (a) the bulk and orientation of some of her component parts, or (b) the lack of bulk and its synthetic creation. The solution of these problems has made possible, in America alone, a 420-million-dollars-a-year industry. To assume that the present corrective equipment is good enough to do the job would presume upon the mental characteristics of the female and the versatility of the industrial designer who never leaves well enough alone.

Among the strangest cases was a vice-president of one of America's largest corporations who came to my apartment unexpectedly one evening, five years or so ago, for a private interview. He had seen us solve a complex design matter for his operating department and we had gained his confidence. So he brought to me another problem: his wife, to whom he had been married thirty-six years, was getting more impossible every day. His young female secretary was getting increasingly adorable. His two sons were in college, his one daughter had just married. Home looked empty, depressing, and hopeless with his termagant spouse. What should he do?

It took me some time to get over the surprise of being put on such a spot. The triangle situation in itself was commonplace, all right, but as a problem it was nothing compared to mine. Its solution was a ticklish one. The problem was how to send my V.-P. back to his wife—where he belonged—without R. L. A. losing the account. I did it in one hour and twenty minutes. He went home, followed my advice, and has lived unhappily ever after. This case history isn't one I am especially proud of from the viewpoint of imaginative thinking. Besides, it put me in the embarrassing position of having broken one of the key rules of our code of ethics, which prevents the designer from doing consultation work without fee.

In 1937 my partners and I felt that the American department store hadn't changed greatly in the past twenty or thirty years. Since the installation of modern elevators and escalators, electric lighting, etc., there had been no major changes in the general philosophy of department store merchandising. Presentation of the articles and the store's physical appearance were pretty well standardized and rather cut-and-dried. We were unanimous in our conclusion that the department store considered as an entity, as a type of selling machine, was ready for redesign. Starting from this unusual approach, we discussed the matter with my good friend, the President of Lord and Taylor, Miss Dorothy Shaver. Dorothy, who is one of the top leaders among America's women executives, agreed with us and promised to give us a chance to demonstrate our ideas in collaboration. She had many brilliant conceptions of her own and we blended our thoughts during the planning of a suburban prototype unit. This new type of store, located about twenty miles outside Manhattan, was labeled "fringe store." The site was purchased in Manhasset, L. I., after a survey made by our office.

A whole new world opened up for my design organization the first day we convinced a client that a store was an implement for merchandising and not a building raised around a series of pushcarts. Proceeding on this premise, we now have one of the largest store-planning units in the country operating within the industrial design organization.

The same type of research, imagination, knowledge of sales techniques, a sense for the dramatic presentation, a feeling for customer preferences made a successful store as it had made a hot-selling refrigerator.

One of our first standout designs was for W. T. Grant in Buffalo. Here we presented what was then a radical merchandising technique in the form of a daylight selling window. It was my belief (after observing women being drawn irresistibly to sales counters where other women clawed each other to buy slightly reduced, mightily mangled merchandise) that nothing is so attractive to a potential customer as the sight of other people buying. I call it "contagious buying."

The daylight selling window is a semicircular projecting window in which we placed a counter carrying small items that sell rapidly. Passers-by, observing a group of women waiting around this counter for service, are literally dragged in off the streets to join the happy throng of women waiting for objects no more exciting than a lipstick or a package of hairpins. Once inside the store, the customer is considered just a cough above a dead pigeon. There he is subjected to the blandishments of expert salespeople, attractively lighted and decorated departments, and the whole dazzling vista of possible possessions.

But there is a whole new philosophy of design for each store, service station, specialty shop, or whatever. The daylight selling window was fine for a store selling low-priced items in bulk to customers demanding little or no service. Lord and Taylor, however, for which New York store we have done modernization of their downtown store and three completely new suburban stores, is almost a way of life in itself. Dorothy Shaver said that there was a Lord and Taylor customer, as distinct from any other customer as a lady wrestler from a 1935 debutante. This 1935 debutante and all her sisters, cousins, and aunts-to-come inspired the almost loving personal solicitude in which Lord and Taylor stores envelop their customers. We created Little Shops—gay as country fairs and as cozy as possible—wandering around the peripheries of informally shaped areas that remove from the layout all traces of stiffness and formality. Counters are gracefully curved, lights are designed to flatter the sandiest Canasta complexion, and the suburban store itself becomes American Suburbia's favorite village green.

Our most ambitious project to date—and the one that seems to point the general direction of our future work in this field—has been the enormous Foley Brothers store in Houston, Texas. This was the first major store to be built in this country in fifteen years, and the techniques developed for this special problem were, many of them, unprecedented. My partner, William Snaith, who directs this division of our enterprise, clarified the problem of all department stores in saying:

1. That the department store is probably the last of the manually operated large industries.

2. That, owing to the personality of its operation, it is difficult although not impossible to mechanize.

3. That when it is caught in the inexorable wage-and-hour pressure, its high cost of doing business gives it very little room to meet competitive price pressures.

4. That operations have been impeded rather than helped by the buildings in most cases, and that improvement could not be applied with a paint brush but would necessitate some fundamental changes in the plants before much value could be felt.

Foley's, then, became one of the first stores in this country to

THE PERFECT FUNCTIONAL SHAPE: THE EGG. Illustrating the skin-tension theory of structural engineering, it is a marvel of design. In spite of its thin shell (7/1000 of an inch), it can support a gradually applied pressure of about twenty pounds without breaking. It is so formed as to create a minimum of friction while progressing through the beast; a good example of streamlining adapted to a slow-moving object. Any other shape — a square, for instance — would make the hen's life intolerable.

Arc Welders. ***Press Association, Inc. Photo.***

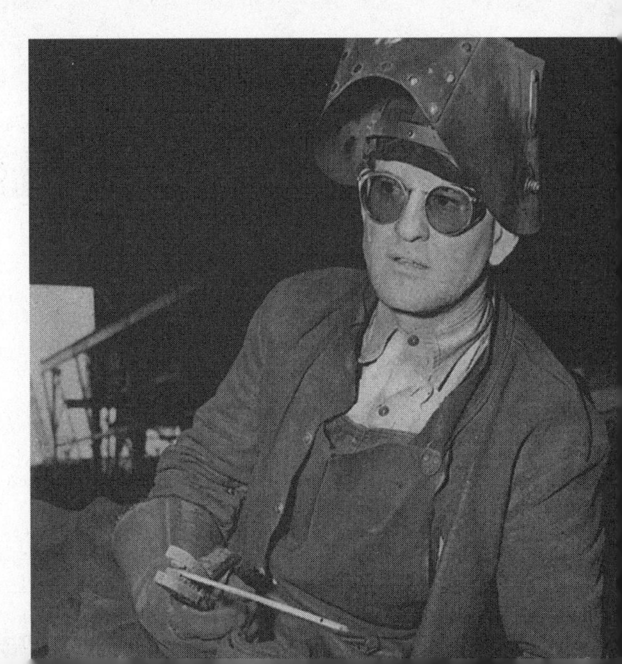

Welder. ***Wide World Photo.***

Welderessed. ***Press Association, Inc. Photo.***

Coal Miner. ***Press Association, Inc. Photo.***

Partners Loewy, Snaith, Bienfait, Barnhart, and Breen. ***Look Magazine Photo.***

Michael and Venise Havinden with the Loewys at Sands Point. *Victor de Palma, Pix, Inc. Photo.*

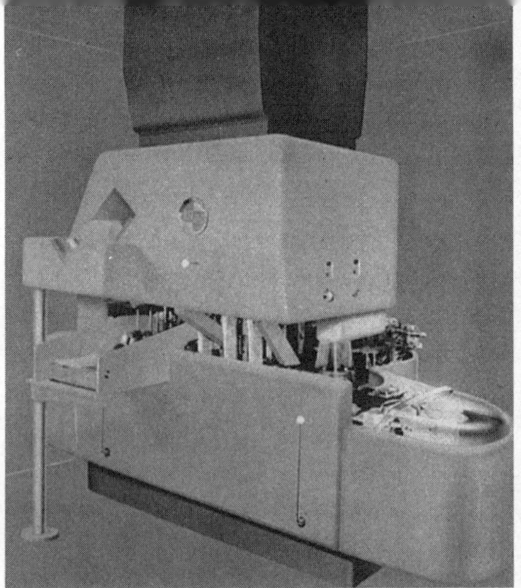

FLEXVAC VACUUM-SEAL PACKAGING MACHINE: The complicated machine on the left has been made dust-proof, more silent and far easier to maintain, as seen at the right.

OVER 30,000 COCA-COLA SODA FOUNTAIN DISPENSERS HAVE BEEN BUILT. Soda fountain operators object to pieces of equipment that occupy valuable counter space. The base of the new unit (right) is more compact. Sticky syrup does not cling to its smooth, flowing surfaces. *Robert E. Coates Photo.*

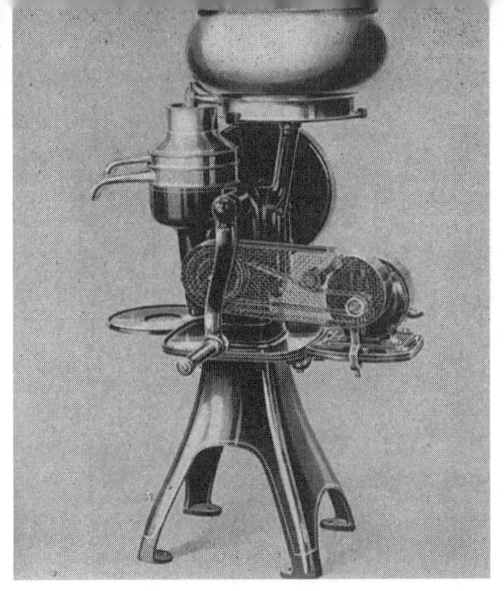

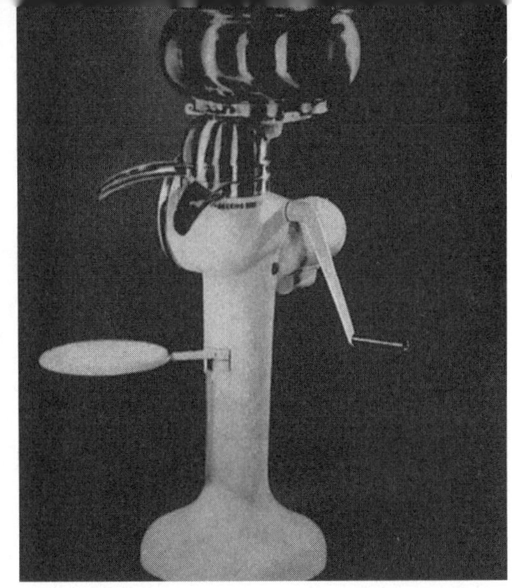

INTERNATIONAL HARVESTER'S CREAM SEPARATOR: The unit on the left was difficult to keep clean. Spilled milk would collect in places hard to reach, making the separator less sanitary.

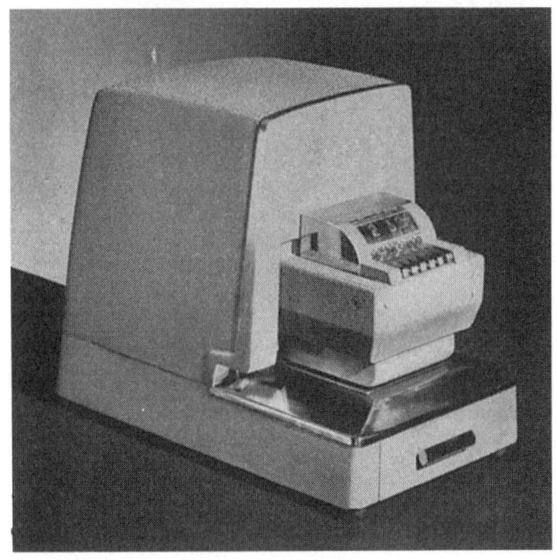

CUMMINS' CHECK PERFORATING MACHINE: The original unit (left) was a foot-pedal operated floor model. Redesign, involving an electronic mechanism, produced the simple, compact table unit, free of dust-gathering cracks and corners, at the right.

Beauty of style and harmony and grace and good rhythm depend on simplicity.

—Plato

Robert Maillart, the famous Swiss engineer, has produced some of the most exquisite bridges in man's history. The above, built over the Thur in Switzerland in 1933, should be brought to the attention of some contemporary designers. Its thoroughbred slenderness is in violent contrast to the vulgarity of their "jelly mold" design approach. *Museum of Modern Art Photo.*

FOLEY'S IN HOUSTON: **The first large postwar mechanized department store.** ***Ezra Stoller: Pictorial Services.***

Mechanized stock manipulation. ***Bob Bailey Photo.***

New parabolic bookshelves improve readability of the books' titles. *Paul Peters Photo.*

Packages are shown passing from one conveyor belt to another at right angles. *Bob Bailey Photo.*

All packages converge upon this sorting station in the garage basement. *Bob Bailey Photo.*

These racks combine display and stock reserve. ***Paul Peters Photo.***

Shoe department, showing perforated display panels. ***Paul Peters Photo.***

Décor blends classic and contemporary styles. *Dorsey and Peters Photo.*

Dress Department. *Gottscho-Schleisner Photo.*

LORD & TAYLOR: Millburn, N. J., Suburban Store. ***Gottscho-Schleisner Photo.***

INTERNATIONAL HARVESTER STORES AND SERVICES CENTERS: To date, 851 have been built; another 394 will be completed this year. (Planned and designed by Raymond Loewy Associates.) The tower is INTERNATIONAL HARVESTER red with black and red trademark. Another 160 are being erected in twelve countries. *Hedrich-Blessing Photo.*

The floor layout and lighting are simple, economical and effective. *Hedrich-Blessing Photo.*

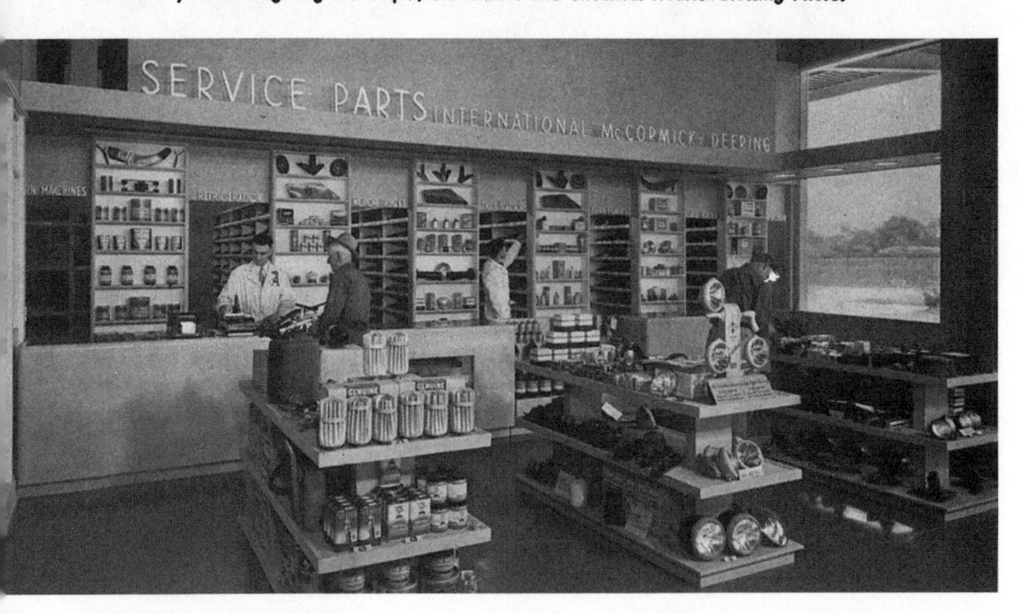

Parts packaging (also by Raymond Loewy Associates) facilitates classification in behind-counter storage shelves. *Hedrich-Blessing.*

A LUCKY STORES SUPERMARKET IN CALIFORNIA: Latest type of supermarket designed for easy shopping, easy maintenance, and highly mechanized stock manipulation. There is plenty of free parking space. Exterior color scheme is lemon yellow and lime green. *Stone & Steccati Photo.*

Shopper's rubber-tired cart has a directory directly in front of customer's eyes. Carts can be folded flat and stored. Departments are identified by numbers. ***Stone & Steccati Photo.***

Floors can be washed easily on account of the cantilevered base of the merchandise racks. Notice slanted shelves for better visibility and stability. *Stone & Steccati Photo.*

Merchandise is clearly displayed. The top label of the packages is reflected in slanted mirror above. *Stone & Steccati Photo.*

Stocking up takes place after closing hours. Labor-saving equipment is used extensively. Notice the ten large removable containers on the truck. *Stone & Steccati Photo.*

Containers are unloaded from truck. *Stone & Steccati Photo.*

Shelves are stacked from pre-loaded containers. *Stone & Steccati Photo.*

make major use of mechanization in order to operate on a sound economical basis. Several devices were used: all merchandise and stock are adjacent to their departments (except for floor coverings, furniture, major appliances) to reduce the number of handlings. Thus the "windowless" building emerged from the practical consideration of how best to allocate space, and not vice versa. All the ensuing publicity about the "windowless" store made a feature of something which was not done for effect but resulted from very sane planning; mechanical aids were installed, among them a wheeler-lift device which keeps replacement merchandise flowing into the store without confusing the selling function. Other mechanical aids are the chutes from selling counters to a conveyor system, thence to a sorting ring, thence underground, to the customers' garage, where packages are delivered to the outgoing cars. Customers are encouraged to do their own selecting, for which purpose two new types of fixture were designed and installed. The entire layout is designed to make it easy for a customer to buy a complete wardrobe in one size and price range without wandering all over the store. Vertical traffic has been given careful study, and escalators and scientifically placed aisles direct the flow.

Beyond all this there is the charm, color, light, beauty, comfort which the American shopper has come to expect. In retrospect it is amazing to see that these factors were the only criteria in our first designs for stores. Today we are members of the selling team.

Our complete modernization of the venerable and frenetic Gimbel's in New York graduates us forever from the decorator class. Mr. Gimbel is perfectly delighted to go along with a crystal chandelier (now that he's relinquished the habit of thinking of the store as a slightly moth-eaten overstuffed chair) if we can prove—through his accounting departments—that the "profit picture" is twice as handsome. I find that a whole new language has come into the office—"profit performances," "expense control," "closed storage systems." Whatever they all mean, the result has been better design, better sales, happy clients, much wampum.

Gimbel's, I love you.

In two important respects the Gimbel and Foley designs were similar. Each required the most painstaking study before the design and architectural phase, and each is a store which judges its success by volume business.

Foley's was new from the ground up, but it is headed by one of the country's smartest merchandisers, a man for whom anything less than stunning performance is unthinkable.

Preliminary study was based on achieving his ideal for a department store with as much stock as possible adjacent to the department selling it and maximum mechanization to reduce operating costs. Service in a store is one of its most expensive functions—alterations, returns, the charge account, personal sales service. Wherever it was possible to use a device it was installed.

As a consequence the vascular system of the store is a combination of chutes, conveyors, automatic stock elevators, escalators, and service companionways ringing each floor. The car in the stock elevators is itself a little truck which rolls out into the behind-scenes stock area where it can be unloaded rapidly and returned to the basement for more of same. If a purchase is made on the upper floors it is wrapped and chucked into a spiral chute feeding into a central conveyor system. Main-floor packages are also popped into the conveyor from behind-counter stations. Eventually this moving belt ducks under the street through a tunnel leading into the garage building. Here, at a sorting ring, several men unscramble the packages, routing them to trucks for delivery or to shoppers in automobiles on their way out of the garage.

The store is windowless as the result of the peripheral stock arrangement and not because it seemed the most dramatic store-building treatment. That the latter results is evidence that function can produce interesting new architectural forms. (Windowless stores are not new, of course, but few can boast that they got that way because of something planned inside instead of a look that was wanted outside.)

Gimbel's study—department by department, square foot by square foot—was the most detailed we had completed up to that time. Another factor entered this design plan: the store had to maintain business as usual despite painters, builders, moving days, et al. A profit picture was established as a goal for each department. Planning and design from then on were based upon one thing only—the realization of that goal. Every design detail had to be justified on a dollar-and-cents basis.

Departments in the store were rearranged, shifted, or eliminated when a profit advantage could be proved. Again management established a purpose for the design, which was the guiding philosophy for the entire design: Gimbel's had attained its status as one of the country's ten largest stores during the war, and surely one of the outstanding promotional stores of all times. Its future was largely dependent on a twofold program—to maintain its heavy volume and type of promotional sale while building a stable customer group in full lines of staple, nationally advertised goods.

Size, size! The problem of design was always size at Gimbel's. Floor areas figured in acreage. The bazaar—the "hundred-million-dollar-a-year pushcart"—has been converted into a series of floors with permanent peripheral shops supporting a center area given over to the dearly loved bargain counter.

Things can be done for the personnel too. For instance, we felt that the soda fountain operators at a large chain store were poorly uniformed and we suggested gold buttons. Their tips rose from sixty cents to two dollars a day.

CHAPTER 17

☆☆☆☆☆☆☆☆☆☆ ☆☆☆☆☆☆☆☆☆☆

Schmaltz, in the industrial design profession, is a widely used term that connotes everything that is vulgar, ornery and superfluous. It may be expressed by an overabundance of gadgety decorations, stripings or curlicues. It is generally accompanied by a cheap color sc

HEME OF CLASH-ING TONES AND CRUDE VALUES. O

ne sure way to commit Schmaltz is to overcrowd a product with disconnected "decorative" elements. In this typographical demonstration, both type and arrangement of copy are definitely Schmaltz.

☆ ☆ ☆ ☆ ☆ ☆ ☆

THE CHROME AND YOU

In the premachine age products were made by craftsmen: artists who devoted their lifetime to making one type of thing, in one medium, freely, and in a leisurely way. They knew their material intimately, whether it be glass, clay, or pewter, used it skillfully and without inhibitions. Design was a hit-or-miss process. It could afford to be so as materials were inexpensive or reclaimable. Molds, if any, were cheap, and a variety of models could be made at little cost. The good ones were praised and preserved, the poor ones discarded and soon forgotten. For many a *chef d'oeuvre*, there was many a failure. For each delightful signed Hepplewhite, Revere, or Cellini there were plenty of clumsy attempts, now vanished. Happy times when the designer could afford to make human mistakes, forget them, and live in peace.

But things have changed. With the advent of mass production, where the manufacture of one single product may call for dies and tools costing millions of dollars, the designer can't afford to be casual in his approach. Besides the enormous capital involved, there is the future of the company to be considered and, therefore, of all those employees whose living is directly affected by the success or failure of the product. If his conception is faulty he will be plagued and harassed day and night by eight hundred thousand shrieking mass-produced monsters that will drive him to a nervous breakdown and ruin his reputation. A far cry from the craftsman's responsibility—responsibility to himself and none else, as he worked alone.

The industrial designer never works alone. As a rule, he is one of a trio (backed by dozens of assistants) whose duty it is to solve the design problem. He, the engineer, and the cost analyst constitute the force whose task it is to consider every minute phase of the undertaking and to solve it with speed, thoroughness, and finality. They, as a team, accomplish in the mass-production field what the craftsman used to do alone with his hands and without outside interference. In his case, the intimate relationship between matter (whether clay, glass, or pewter) and artist infused in the finished product the true expression of his own personality. He had absolute control over his creation from start to finish. In some cases, the object did not even need to be signed in order to betray its authorship. One could tell at a glance and a feel, "This is a Stradivarius, or a Phyfe, or a Della Robbia." Contrariwise, a mass-manufactured product needs to be perfected from first sketch to completion by a team of men and in a welter of unavoidable interferences such as suggestions, countersuggestions, criticisms, and technical restrictions. Countless stages and situations threaten the design integrity of the product from all quarters and menace its very existence without warnings. To protect his infant during this tumultuous northwest passage across the wilderness of research labs, mock-up rooms, model shops, conference boards, and testing labs, the designer must be gifted with the vigilance of a trapper, the diplomacy of a Talleyrand, and the perseverance of a Columbus. The dangers are constant and lethal, as the tendency of things, if left alone, is to take with uncanny sureness the wrong direction. Man has long ago discovered the "perversity of things."

Without going into the nature of these constant threats to the very life of the design, let us say that to me it is a fascinating expedition. When the talented members of such a task force are men of taste and determination, with mutual respect and a sense of humor, the pace is breath-taking and it is time for the competition to feel sorry. The tactical maneuvering of such a design force, once the strategy has been established, equals in precision a by-play by Notre Dame or the brilliance of a great surgical team in action. There you can see lucidity, economy of action, continuity of thought, and executive courage at its peak. It is an exciting experience and one that could take place nowhere but in America.

I can remember examples of such happy teamwork with many a client of ours. Sometimes, as in the case of Frigidaire, the design process is such a thrilling one that it becomes fascinating. Our clients' requests for business trips to Dayton become events that we are waiting for. As a result of such joint efforts, the handsome product shows clearly that it hasn't suffered at birth, and its proud makers lead the whole industry in both sales and prestige year after year. For the benefit of the reader who might be interested in a typical day with the design task force of a corporation, we shall have a blow-by-blow description later. In such situations, where everyone concerned tries to understand and help the designer, the final product is a true expression of their taste and talent. It is a great satisfaction to see that the dominant characteristics of design have remained intact up to the production line. In such a case, the designer can walk incognito among the thousands of dealers and distributors at the ceremony marking the introduction of the model and hear such remarks as, "You can sure tell who designed this stuff!"

Much has been written in recent years about function and aesthetics. The usual theory is that "anything functionally correct is bound to be correct in appearance," or "if it works well it looks good." This statement needs clarification as it isn't necessarily true, especially in the case of complicated machinery. However, it is usually applicable to simple objects. For instance, almost everyone takes it for granted that many of man's simpler products have reached perfection in design. The ox helve, the plowshare, the ship propeller, the needle, the glass retort are functionally correct and aesthetically harmonious. It might not be true, however, in the case of more complex units such as a threshing machine, a cotton picker, or a textile loom. These are satisfactory, functionally speaking, but their appearance is messy and disorganized, the ensemble is disturbing. In other words, function alone does not necessarily generate beauty and it seems that there can be no beauty without order. The harvesting machine may be excellent in operation, each component part may be perfectly designed in itself, but in the ensemble the result is bad. Why? Because the machine, as a whole and regardless of its well-designed components parts, is and looks complicated. So here, I believe, is a true answer to an industrial designer's theory of aesthetics: It would seem that, more than function itself, simplicity is the deciding factor in the aesthetic equation. One might call the process: beauty through function *and* simplification. Therefore, the test of the designer's ability lies in his success or failure in establishing simplicity through order. To achieve this, the avenues are clear. First, each component part must be designed efficiently with utmost regard for economy of material. Second, multiplicity being the essence of confusion, the designer will endeavor to eliminate or combine parts, supports, or excrescences whenever possible. This technique I would call "reduction to essentials." He will then take into consideration colors, textures, and finishes as well as materials themselves and submit them to the same simplification treatment. When every component part has been stripped down to its simplest form, every duplication ruled out, projections and asperities reduced or eliminated, colors and textures simplified, the result is bound to be aesthetically correct. I claim that functionalism alone could not have achieved such a result. For the sake of the record and to avoid possibilities of misunderstanding and misquotation, let me state once more that I am all in favor of the "beauty through function" theory. However, I would like to alter it to "beauty through function *and* simplification."

Besides this tribute that design pays to aesthetics, it pays in some other ways too. It is good business, as it sells the product. Let us examine the reasons why.

A product can be engineered to perfection, manufactured with precision, priced right, function well, and still be shunned by the public. The reason is that the buyer has formed in his mind a preconceived idea of what a "fine" product of a given kind should look like. Any deviation from this will prejudice his appraisal when he comes to go through the cold ordeal of parting

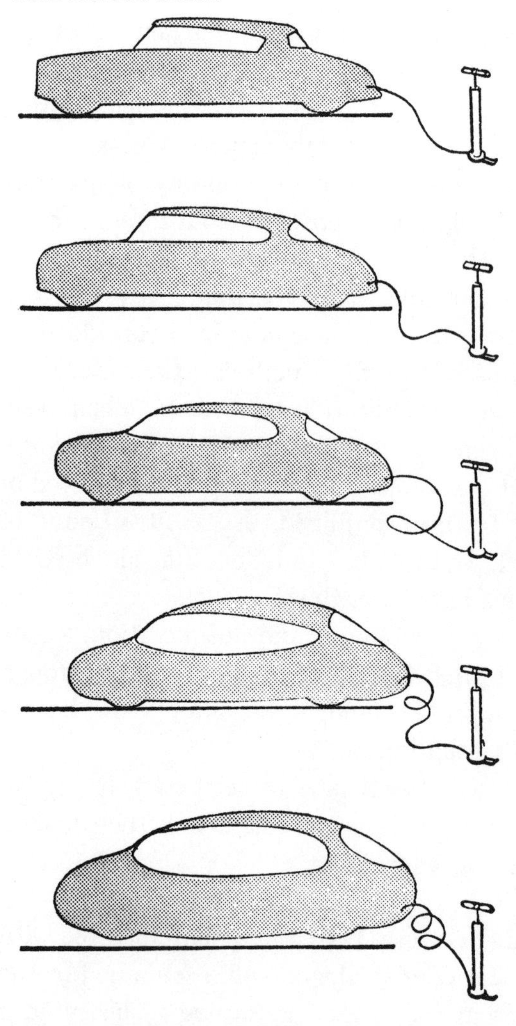

THE JELLY MOLD SCHOOL OF DESIGN

OR

THE INFLATIONARY TREND

with his cash. In other words, the finest product will not sell unless the buyer is convinced that it is really the finest.

The competent industrial design operator knows what constitutes the consumer's picture of a fine product. He has a knowledge of the factors that are repellent to his taste or appeal to it. It is the designer's duty to eliminate the former and add the latter. At R. L. A. we call this process the up-grading design system. Without getting into the technical aspects of the subject, let me say that the process of simplification described above encompasses briefly its major points. To these, we may add the elimination of cracks, joints, rivets, screws, etc. Another thing to be avoided is the feeling of boxiness. The general mass should be flowing, graceful, free of sharp corners and brutal radii. This within the bounds of restraint, of course, as many a designer overdoes it and the end result is bulbous, fat, and without character: what we call the "jelly mold school of design."

Let us consider, for the sake of illustration, a typical up-grading design job. Toaster A is the largest seller and rated A-1 in every respect. It is sleek, silent in operation, highly polished, and smoothly finished in every detail. The base fits neatly and the whole constitutes an integrated unit. The side controls blend well with the shell. There are no rivets or joints visible. The price is $18.95.

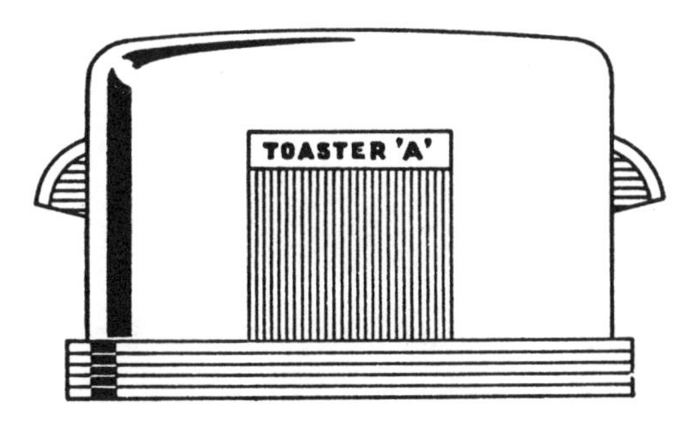

Toaster B is made by a competitor. Quality and performance are equal to A but it does not sell as well. The shell shows several joints and a few rivets. It is boxy-looking and the edges are sharp. The base does not harmonize with the whole; neither do the control knobs. The price is also $18.95.

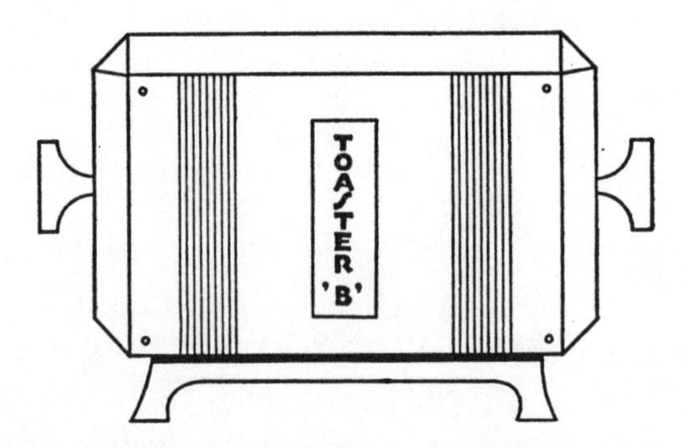

The industrial designer is called in consultation. He examines sick toaster B. The diagnosis is boxiness with a touch of angular deformation. Also a mild case of rivets and knobby protuberances. Nothing wrong with the insides, and functionally everything is okay.

The prescription is simple. Just a face-lift job. Trouble is that its owners are very impressed by toaster A and they want their product to look like it. "The public has made up its mind," they say, "that it looks like quality and anything that looks different they won't buy."

The designer is on the spot, because he absolutely refuses to copy toaster A (or any other toaster), but does not say so to the client. For he knows that he can do better than A and that when they see his design they will have to admit that it has all the quality feel of A and still is original. So he goes to work and he isolates in his mind the features of A that attract the consumer, the ones that repel him. Then he produces toaster B1.

Toaster B1, unlike B, has no sharp edges. Rivets are gone; so are the cracks and offsetting surfaces. The plain base blends even better with the shell than it does on toaster A. All three are of the same height, but the thin black base and the horizontal black controls make B1 look lower. Qualitywise, it looks even more convincing than A. The mechanism has been sound-insulated. The price is $18.95.

This product has been up-graded and the chances are that it will sell well, far better than its predecessor.

Oversimplification also has its dangers. It never reaches the border of vulgarity as does its counterpart, the "jelly mold job," but it gives the design an arid look and, often, a feeling of unbalance. The wheel, this magnificent device, has been subjected to all sorts of design treatments more or less successfully. Let us see how it has fared since the early days of the "machine age."

THE INDUSTRIAL WHEEL SHOWS CLEARLY HOW DESIGN SIMPLICITY CA

Quaint and Silly

Superfluous

Cute

IMPROVE A PRODUCT AND HOW OVERSIMPLIFICATION CAN HURT IT.

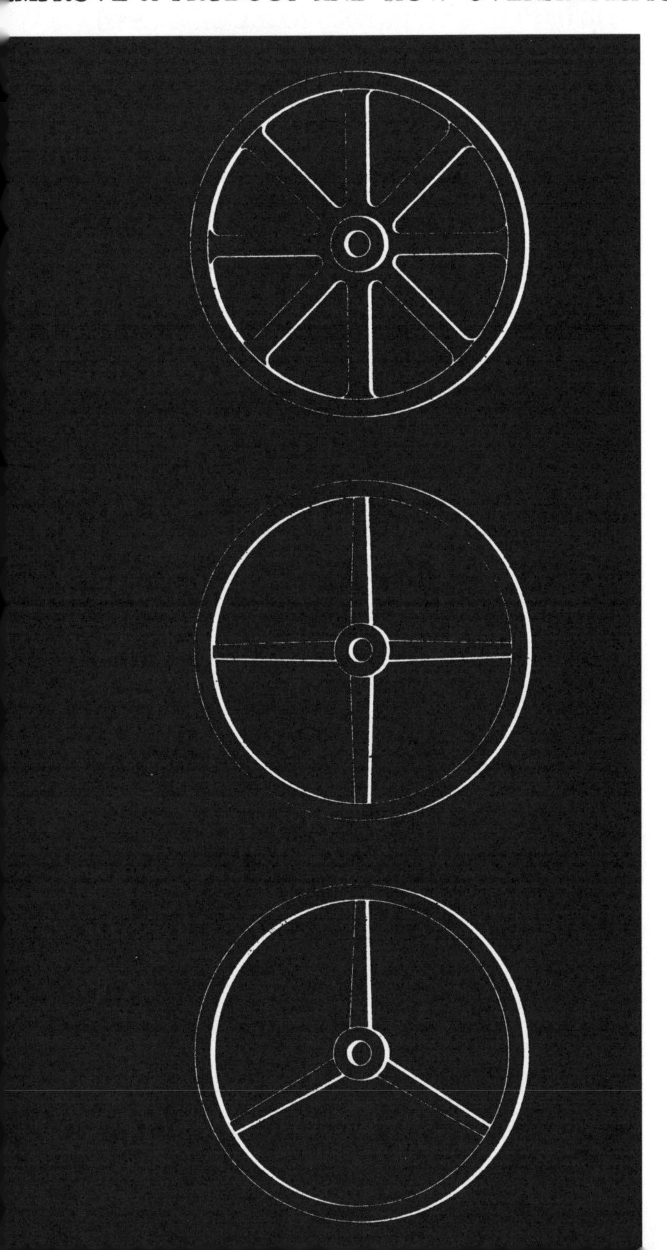

Simple and Dull

Neat

Perfect—but can it be further simplified? No, not successfully. (See next page.)

This wheel, reduced to its simplest expression seems unbalanced and generally frustrating.

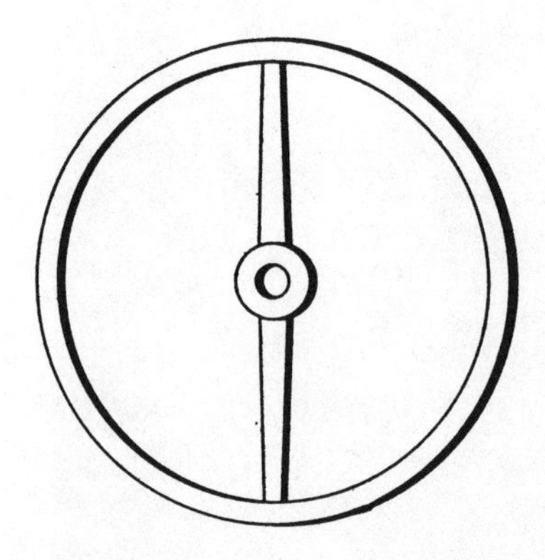

Therefore, let us settle for this: *

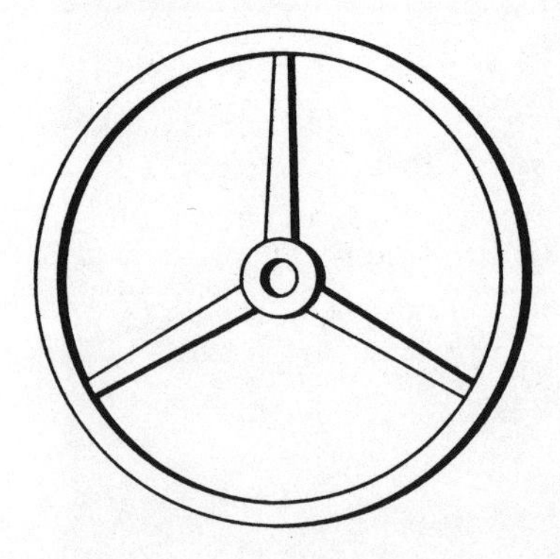

* The reader may say, "Does he mean to leave well enough alone?" Not on your life. There are dozens of other ways to improve it further, but outside of the scope of this book. R.L.

Sometimes a product is so complex that design simplicity cannot be achieved, as in the case of a modern airliner, for instance. Anyone who has visited an aircraft plant and who has seen the makings of such an aircraft, with its hundreds of thousands of parts, gadgets, wiring, tubes, dials, ducts, spars, levers, rivets, and controls, must admit that it looks extremely confusing. As modern technology cannot simplify it much and keep it operative, we have reached an impasse. For the sake of argument, if the skin of such a finished machine were to be transparent plastic, instead of opaque aluminum, the aircraft instead of a slender form of great elegance would appear to be a nightmare of confused engineering. Yet, it would still be functionally correct. The aluminum skin saves the day and transforms it into a sleek masterpiece. Another example is the hood of a motorcar. If it were transparent, instead of a neat vehicle we would see a messy-looking mass of complex machinery in apparent disorder. In other words, there are cases in which a shell or wrapper treatment is in order and justified. This is contrary to the conceptions of a great many aesthetic critics of the industrial design profession who have branded the "shell technique" by all sorts of ugly names. These purists call it a breach of design integrity. To this I can't entirely subscribe. I believe that when a given product has been reduced to its functional best and still looks disorganized and ugly, a plain, simple shield, easily removable, is aesthetically justified. This shield itself accomplishes something, and it becomes functional, the specific function being to eliminate confusion.

On the other hand, the designer who resorts to such a device without having assured himself that there is no straightforward design solution is guilty of professional carelessness. Superficial critics often cite the animal kingdom as an example of beauty through function. How beautiful a shark can be! What a magnificent creature built for speed and maneuverability! I agree, a shark is a first-rate design. But what about the giraffe? It is just as functionally designed for speedy action across the underbrush, and its long neck provides an ideal lookout for enemy detection. Yet the theory of beauty through function has not worked in this case. The giraffe is ugly in spite of its functional excellence. In order to give my critics a break, I must admit that my "simplification theory" would fail, in this case, too. Even if the legs were reduced to three and the tail cut off, the poor beast would still look ridiculous.

The animal world seems to substantiate the acceptability of a shield or skin covering. In the case of Miss Betty Grable, for instance, whose liver and kidneys are no doubt adorable, I would rather have her with skin than without. Furthermore, it can be argued that Miss Grable's skin is really not a shield as such but a functional unit that serves the definite purpose of generating beauty, and therefore is desirable in itself.

This shield, or housing, reminds me of a conversation I had, years ago, when streamlining was news. One could hear such expressions as "airstream," "airfoil," "teardrop design," etc., used and abused ad nauseam. There were streamlined shoe polishes, airfoil wallets, and teardrop meat grinders. At the time I was in contact with a manufacturer of small electrical equipment. He was making an honest hair dryer that did a good job of hair drying. All it needed was a better arrangement of its component parts, a sturdy switch, and a new trademark. But our friend, who was a bit of a diamond in the rough, wanted to go further and louder.

"Say, Loewy, lookahere. It ain't enough to clean up the darn mess. I needa lotta sparkle and shine. Ya don't get the idea, I'm tellin ya."

"But, Mr. Blurg, if you will let me explain . . ."

"Now, looka here. What I want is class, see? You putta the doggone mess in a shell, see, a shell like a big long egg, whaddya calla the shape? The streamline shape? Oh yeah, a gumdrop design, and I wanta lotta chrome! Kind of a big chrome-plated gumdrop!"

Those days are over but the chrome angle is still a major factor, and a major headache to the discriminating designer. Our theory about chrome, or about any kind of bright metallic finish in general, is that a certain moderate amount is desirable as an accent, as the subtle touch that will, if correctly located, add a pleasant spark of highlight to a well-shaped surface. When splashed all over without either rhyme or justification, it is poison. It is vulgar, gaudy, and an admission of the designer's lack of taste and design imagination.

But it has been the designer's sad experience that a great majority of the people love chrome, and love it indiscriminately. Often, I have designed a product with basically correct forms so that its own highlights and shadows would inherently create the right sparkle and make chrome decorations unnecessary. In most cases, it has proven insufficient. As soon as the product was produced and placed in the field, the consumers requested more chrome. This has been especially true in the automobile field. The buyers pile up their own chrome-plated gadgets, moldings, strips, ornaments, etc., until the car looks like the well-known Christmas tree. Part suppliers make and sell tens of millions of dollars' worth of inept and ugly "bumperettes," hood emblems, bumper guards, sun visors, etc., that our consumer friends plaster all over the car. So now, in self-defense, the designer incorporates in the design more chrome than he would normally choose, but at least he can control its distribution over the surface and avoid monstrosities.

However, there is always some ivory tower modern art critic to take up the holy crusade against the poor designer, accusing him of cheapness, mass vulgarity, and aesthetic prostitution unbecoming the conduct of a professional man. What I would like to know is what these critics would like the designer to do. For him to train and educate the masses in aesthetic appreciation of simple, beautiful form will take several decades. It is a proven fact that there is as yet no general public acceptance of products whose design has been reduced to their simplest expression, outside of a limited segment of sophisticated buyers, representing perhaps a few per cent of the consuming public. So again, what is the designer supposed to do? Design for his client a product that will not sell, that may put the company on the rocks, create tens of thousands of unemployed? Isn't it preferable, for everyone's benefit, to make this educational effort a progressive one, to wean the public away from chrome through a subtle but constant process? And so doing, keep the manufacturer busy and sustain employment? My clients, my associates, and I agree, and I can promise the purists—of which I am one—that the goal is consistently before our eyes. Only please give us a chance to reach it in an orderly fashion.

In the meantime, and as a welcome relief, I occasionally design a product for my own use—whether a town car, a boat, or a radio set—reduced to essentials and with practically no chrome at all. This is an affirmation of my belief that a well-shaped form with intrinsic elegance of line and economy of material needs but a slight occasional dash of bright finish to accent a highlight and darken a shadow.

All executives are not like my chrome-plated-gumdrop friend. Many have a keen sense of aesthetics and are conscious of their responsibilities to their firm and to the public. One outstanding example is the vice-president of a company manufacturing fountain pens. Contrary to the rest of the management of the company, and in agreement with me, he was violently opposed to the adoption of the ball point pen, in which he had no confidence. The story is that one day, after a long and unsuccessful struggle, he was completely fed up with the whole business and ready to quit rather than okay the new pen. So he called his secretary, dictated his letter of resignation, prepared to sign it—and the pen refused to write. He reconsidered, stayed with the company, and has been miserable ever since.

Even the best-planned and designed product, when placed in operation, may divulge certain defects or shortcomings that need correction. This is especially so in the field of transportation. For instance, when new types of folding washbasins were used in Pullman cars, some passengers forgot their dentures and dumped them on the tracks with alarming regularity. There were a great many lisping Chairmen of the Board who presided at board meetings in Chicago with their teeth between Gary and Englewood. This had to be stopped. Stopped with a strainer.

About 1940, new types of de luxe coach trains were put in operation between New York and Florida. The "consist" of these trains included the latest type of coaches, with reclining seats that were very comfortable. The armrest between seats could be folded up, and when two seats were fully reclined they became, more or less, a double bed. After a while, it turned out that many young couples had discovered this characteristic. What with a nice dinner,* a few drinks, the perspective of sunny Florida, the lights off, and the friendly co-operation of a fur coat spread out fully . . . !

Before long, the railroad found in its mail numerous letters from shocked passengers who objected to this unadvertised form of travel comfort, and thereby paid unwilling tribute to American manhood. So we had to install an electric light bulb under every other seat, a type that could not be turned off by the amorous passengers.

Just before the war our London office designed a couple of trains for the state railway of a rather primitive country in the Near East. In a few weeks, anything that could be unscrewed, unhooked, or unbolted had disappeared. Cars were left with no roller shades, no lights, no seats, no baggage racks, and no door handles. The passengers would throw these articles through the windows to business associates who were waiting along the tracks. It was very costly to the company, in spite of what could be said for this distribution of souvenirs from a public relations viewpoint.

In the next trains we designed, everything was welded instead of bolted; seat cushions were eliminated. It stopped pilferage, as few passengers carried welding equipment in their luggage.

As an experiment and in order to reduce weight, footrests were

* Shrimp cocktail, creamed chicken with peas and carrots, ice cream, and coffee.

removed from a transcontinental bus. Whenever the vehicle would accelerate or decelerate rapidly, an assortment of empty liquor bottles would roll all over the floor. At night it would wake up the passengers. We had to replace the footrests and re-establish peace.

Later, as a test, I specified a plain beige fabric as upholstery for the chairs of another long-distance bus. After a few weeks the chairs looked horrible, full of spots, grease, dirty blotches, etc. We made a survey and interviewed the Chief of Maintenance. He thought that plain fabrics would never be practical, but that a design pattern would blend with the spots and tend to make them less visible. We went further and inquired about the average size of the spots (a fifty-cent piece) and the nature of the spots. They were, in order:

coffee
fat (from sandwiches)
chewing gum
Coca-Cola
milk (for babies)
chocolate
tomato juice
water (from babies)
egg (cooked)
gum

We designed several samples in complicated patterns, in the right size (fifty-cent piece) and proper colors. They were tested, found excellent, and several were adopted for general use. They blend well with most food, except yolk of eggs and tomato juice.

Animal risks are to be anticipated whenever a new plastic or composite material is used. There is, for instance, the case of the progressive young architect who thought ordinary plumbing was too old-fashioned. Instead of running lead or copper pipes under the floor of his dream house, he simply laid soft plastic tubes, like garden hose. It went well until mice tasted it and found it delicious. The constant flood made replacement imperative.

A nuisance that we have to guard against is defacement. One of humanity's strongest urges is the desire to engrave one's name on anything that can be dented, scratched, or embossed. Some go so far as to engrave (or tattoo) their names on their own skins. In transportation it is a major problem. We have to protect bulkheads, pier panels, toilet rooms, etc., from the urge of Joe to sketch a little fresco announcing to the world that he loves Julie. Or that the sitting position is favorable to the thinking process; or that So-and-So is a so-and-so, etc., etc. This calls for the use of certain nondefaceable materials.

Designers are continually preoccupied with safety. We have all heard about the perversity of inanimate objects. If a thing is left to its own devices it will invariably act silly. If you drop your collar button you can bet that it will slowly but deliberately roll across the room, make a right turn around the commode, a left in front of the chair, and, in a sudden lunge, wedge itself into the one-quarter-inch space between the radiator and the wall. The collar button acts exactly as if it had a long-delayed appointment with the dime and the pearl earring that first went there seven years ago.

If left alone people generally act the same way. They will try to insert the square peg in the round hole—or in no hole at all—drop their tiepin into the mayonnaise, the cat in the washing machine, or their watches into the lamb stew, and—mind you—

these are people who are all quite intelligent in every other respect. Unfortunately their behavior is difficult to anticipate.

Accidents happen in the most curious manners. You may remember the case of the Department of Agriculture employee in charge of artificial insemination of cattle. He was driving peaceably somewhere in Indiana when a furious and probably frustrated bull lunged at the car and sent him into the ditch. (Whether or not the bull had a case is open to argument.)

We read all the time about children who swallow safety pins—even baby turtles and penny whistles. Recently they operated on one such little darling and found his stomach full of false eyelashes. There is nothing a mere designer can do about this, short of designing a muzzle for the sweet cherub's little mouth. Even so it is inevitable that some child somewhere would discover that the muzzle would be inflammable, would immediately look for a match, and put two and two together. These occurrences are quite unpredictable.

Or consider the case of the plastic automobile license plates which a Middle Western state (on an economy binge) adopted as standard. Unfortunately, hogs discovered that they were delicious nourishment and they devoured the plates, ciphers and all, with relish.

Bubble gum is our latest problem. Children experiment with injecting it into the strangest places—automobile cigar lighters, television condensers, alarm clocks, etc. They even go so far as to swallow it, for which the designer cannot be held responsible. (It is a distressing thing to consider that "someone" probably designed bubble gum.)

No one needs to be told that safety in the home keeps designers pretty busy. I won't quote figures on the number of serious accidents that are caused by faulty equipment or—more often—by careless use of equipment around the house. Our organization knows most of the reckless acts a human being has thought of to perform in a kitchen, a bath, on a flight of stairs, with electrical appliances or just plain sitting in the living room smoking. It is a credit to American ingenuity that people can think of so many ways to injure themselves. If designers were able to anticipate all of them they could be national heroes.

Several ways of combating dangers in the home come under the designer's responsibility. First, the actual shape, weight, balance, stability—even color—are considered. Handles must be "grippable"; gauges and dials readable; in moving vehicles there must be good visibility under changing light conditions. Sharp corners must be rounded and hinges must be shielded. In the case of the Studebaker, good visibility is credited with having saved many lives.

The human body defines the outline, texture, weight, and temperature of all designs that human beings use. Legs determine seat heights and the depths. Elbows control armrest heights and the padding of surfaces that elbows can hit. Eyes must be protected from glare; they must look through undistorting glass or plastics that will not shatter and fly about.

Design, then, is the first control over the misuse of all equipment operated by and surrounding individuals.

We know that the simpler the product the more beautiful it is usually—and the safer to use. If there are only three gadgets a person can move or adjust on a given utensil or machine there are probably fewer chances for accident than if seven processes are involved. Only a highly trained man can fly a bomber; even a child can open a carton of marshmallows.

Often equipment calls for an actual safety device to make it impossible for the equipment to be used without someone's consciously manipulating a device. The act of stopping a human being for a few seconds before he can initiate the action of the device is as old as time. For this reason we design a lock, a governor, a slow valve, a brake—or we hire a guard.

These devices say: "This thing moves, burns, or cuts, or is heavy; while you prepare to unleash this energy, take warning." If the device is too much for an inexperienced person to cope with we weld, solder, nail, rivet, or bolt the mechanism shut. But then a child or a woman can still outwit a safety engineer twenty to one.

Another control over the misuse of equipment is a set of directions for use. Often we print these on the product in words—simple words. If occasion requires we design in a light that flashes, a bell that rings, a whistle. Failing this, a rule book is provided; we have even designed these to be included with our products.

Lastly, of course, public education in the use of just plain things goes on. Now that we have automatic transmissions on cars and sealed refrigerator compressors, the public is spared the necessity of learning about these complicated assemblies. Automatic features contribute to reducing accident by making it impossible for an amateur to get at them to short their circuits, strip their gears, insert coins, hairpins, or strawberry jam into their most intimate inner workings.

Ideally the definition of the safest product is approximately the same as the definition for the best-designed. It is simple, efficient, has quality, is economical to use, easy to maintain and repair. Luckily, also, it will sell the best and be good-looking.

IN MANY A PRODUCT OF HARD RECTANGULAR SHAPE, SUCH AS AN ELECTRIC RANGE, A REFRIGERATOR, OR SEVERELY CYLINDRICAL ONES SUCH AS WATER HEATERS, IT IS ADVISABLE TO ADD SOME GRACEFUL CURVES AS A DESIGN ACCENT. THIS CAN BE ACHIEVED BY MEANS OF A NAME PLATE, A HANDLE TREATMENT, A CONTROL PANEL, ETC. THE ABOVE SPENCERIAN HEADING CORRECTS THE SEVERITY OF THE TYPESETTING.

Chapter 18

INDUSTRIAL DESIGN AND YOUR LIFE

Mr. Walter Beige, free-lance writer, has decided to do a piece for a monthly magazine about industrial design. He has a "new" slant on the subject.

"Mr. Loewy," he says, "few people realize the scope of your work, and my purpose is to make the reader realize in a very personal manner, *personal*, see what I mean, how your designs affect his life. Yeah, or her life. So here is what I'd like to do: I want to describe a typical day of the average guy from the moment he wakes up until he goes to bed. As we go along, I'll mention all the products, things, or services he uses and that your firm designed, see—the razor, the Frigidaire, his car, etc. We wouldn't miss a thing. We'd list them all, don't you think it'd be a knockout?"

"Well," I said, "it could conceivably become confusing, and the tempo might be slow; also, you might—"

"Oh, I know you are a busy man, but why don't you give me half an hour and help me do it? What about it?"

"Of course, I'd be glad to, but again, I think you'll find that—" He places on the desk a pad and a pencil.

"Oh, c'mon, let's try. Shall we?"

"Sure."

"Okay, here we go. Let's call our fellow Jack Smith. Okay, now Jack Smith is home on a winter morning, it is early, he is asleep."

"Asleep on what?"

"On his bed, of course."

"All right. Then mention that we designed the mattress."

"Okay, so he gets up and—"

"How did he wake up?"

"Well, the alarm clock—"

"Teague designed the alarm clock."

"Right, so he goes to the bathroom . . ."

"In the dark?"

"No, he turns on the lights."

"With what?"

"With the switch, of course."

"We designed the switch."

"Okay, he turns on the switch."

"You say it is winter. Isn't Jack cold in his room?"

"No, because he has turned on the heat."

"What with?"

"Well, with the radiator valve, I guess."

"Then make a note that Arens designed the valve."

"Well, okay. So he walks to the bathroom and—"

"Does he walk to the bathroom on the cold floor?"

"No. After all, he can afford a carpet, can't he?"

"I believe you, but then say that Joseph Platt designed the carpet."

"Good. So he starts shaving with his electric shaver—"

"No, not today."

"Why?"

"Today he is using the new safety razor that he bought yesterday—and which we designed."

"Well, okay, so after he finishes shaving—"

"Excuse me, Mr. Beige, but does Smith shave without shaving cream?"

"No, I guess not."

"Then mention the shaving cream tube that—"

"Okay, okay. So he finishes shaving and—"

"Sorry to interrupt, but you see, he is not familiar with the new razor, and he has nipped his chin. So he reaches for the bottle of antiseptic and a ready-made strip of adhesive gauze; we designed both. And while we are still in the bathroom, and to spare your time, let's say that Dreyfuss designed the tub, the washstand, the shower, that we did the lighting fixtures, his dental cream tube, his toothbrush, the cake of soap, the floor covering, the wax to polish it, the mop, his comb, brush, hair tonic, towels, the—"

"Okay, I get the idea. That'll be enough for the bathroom. So he starts getting dressed, turns on the radio; say, did you design it too?"

"Yes."

"Fine. Now he is dressed, goes to the kitchen for breakfast, and opens the Frigidaire, the designs for which *I know* you made, he switches on the toaster."

"Van Doren made the design for that one."

"Oh, he did, eh? Well, he picks up the coffee percolator—did you make the design for it?"

"We made it."

"The bread knife?"

"We made it."

"Eats his breakfast."

"He made it."

"Kisses his wife good-by."

"We ma— No, sorry."

Our copy writer friend begins to realize that the story may be a bit tedious after all, and gives up the idea. Thank you, thank you, Mr. Beige.

To give a clear picture of the number and types of products we have designed would make monotonous reading. However, it could be said that the average person, leading a normal life, whether in the country, a village, a city, or the metropolis, is bound to be in daily contact with some of the things, services, or structures in which R. L. A. was a party during the design or planning stage. In fact, it is reasonably certain that over 75 per cent of the American population is directly affected at least once a day. It may be on account of a trip on a bus, a note to be written, a smoke, a broadcast, a shave or a snack, a meal, a drink, a shower or a ride, a flight or a kiss.

So in less than twenty-five years, industrial design has grown from a hesitant, unsure probe into what *Time* called "a major phenomenon of U. S. business." The young profession is rapidly forging ahead. It is generally agreed that industrial design is just as important a factor as advertising in the successful marketing of a product, a service, or a store.

The case of the Studebaker car is a good example of what it can do when it is coupled with top-flight management, correct engineering and manufacturing. When R. L. A. was retained in the late thirties, the company was just out of receivership. Since the war, Studebaker has broken all its peacetime records for sales and profits. 1949 was a record year; indications are that 1950 will surpass it by far. At a time when independent manufacturers are struggling to maintain their position, Studebaker is rapidly forging ahead and is now the largest independent—right on the trail of the Big Three.

There are many new fields which industrial design has not entered yet, fields that are badly in need of its services—vast operations and manufacturing realms permeated with ugliness, inefficiency, or archaic backgrounds. Noise is our next target, because noisy operation means poor design. Noise means wasteful use of energy; noise is expensive; noise is a parasite.

There are a few other improvements that I could think of and that would make home life a better life. For instance, the designer could devote his talents to the problem of the American pajama. Pajama pants ought to behave properly. Every man knows that in the magazine or newspaper ads they look like this:

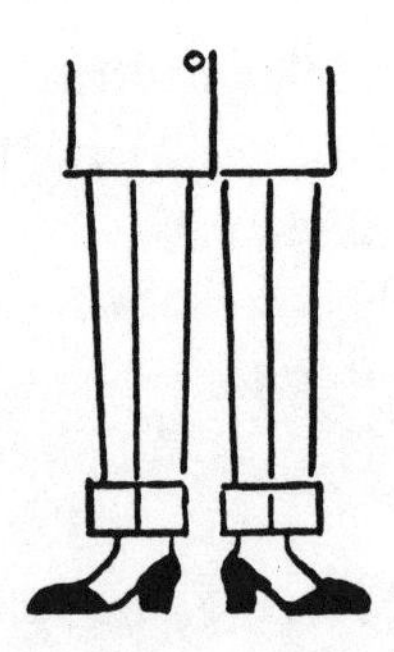

After one trip to the laundry, whatever is left and returned to you looks like that:

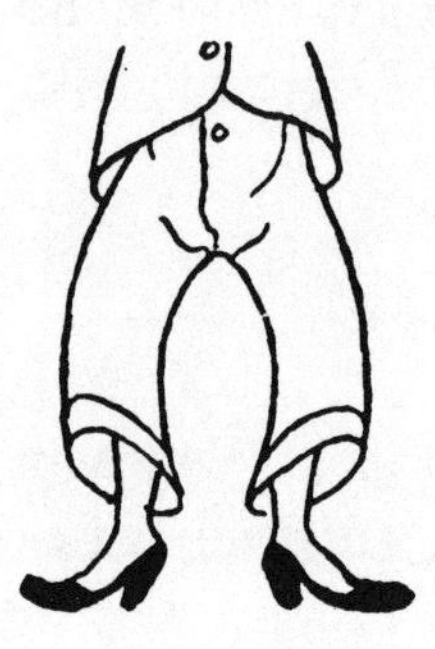

And next morning:

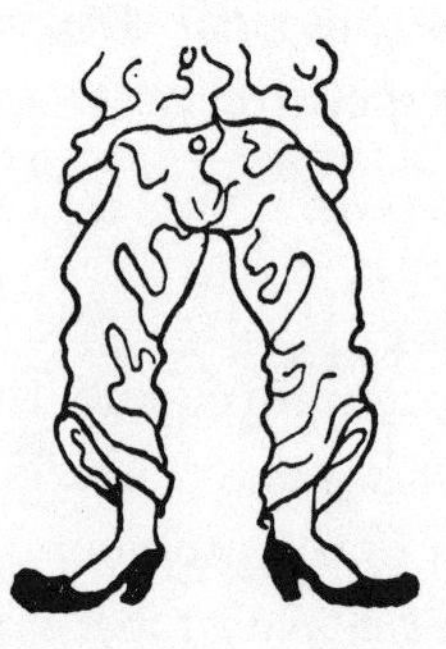

(Which may eventually lead to divorce and suicide.)

Other improvements in man's daily life could be brought about immediately, and they would not even involve the industrial designer's co-operation. To mention a few, I would suggest:

Immediate world-wide destruction of:

1. Cigarette lighters (except Zippo)
2. Zipper fasteners
3. Ball point pens
4. Painted men's neckties

Also, a ban on some offensive practices—things such as:

1. Waiters and waitresses presenting the check upside down (a purely hypocritical gesture)
2. Ugly, dirty, and impractical taxicabs
3. The "uniform" of their drivers
4. Subway filth, noise, drafts, and smell
5. Firms that mail advertising literature in envelopes marked "Personal"
6. Sadistic career women who crush and destroy any talented and attractive young woman that may come their way in the organization
7. The unnecessary discomfort of dentists' chairs
8. Putting sugar in mayonnaise
9. Weather talk by elevator operators

I should like to give an idea of the scope of our activities without going into the lengthy details that killed Mr. Beige's story. For instance, in the transportation field, we have been retained by the Greyhound Bus System for fifteen years as consultants in the design of their prototype busses, the latest being the Scenicruiser, with its observation turret. In the same field we are just completing our design work for the three new American President Line ships. And, naturally, the next Studebaker passenger cars and trucks. These are but a few examples at random.

In the products field, we are now developing for Frigidaire the next models of refrigerators, electric ranges, freezers, air-conditioning equipment, and various household appliances; also, hundreds of kitchen utensils for Ekco products and a line of modern china dinnerware to be made by Easterling.

For others, we are designing a new type of bicycle, whipped cream dispensers, cigarette lighters, and outdoor grills. Problems in a different vein are a new type of dictionary, psychological test cards for the University of California, an exhibit hall for the state of California in the city of Los Angeles, and new branches for one of the major banking firms.

More than thirty new stores are on the drawing boards, among them some very large ones, and others small, such as a job on the famous King ranch in Texas. Also railroad stations and operating-room equipment.

Again, in the products division, let me mention storage batteries, filling station equipment, new methods for the dispensing of drugs, precision scales, cameras, kitchenware, and dozens of other items.

In the packaging field we take care of the design requirements of Armour and Company, Lever Brothers, National Dairy Products, National Distillers, American Tobacco Company, and twenty other corporations.

These are a few examples among our 140 clients, all active. Our employees number approximately 150.

We are about to move our office to a new building just being completed at 488 Madison Avenue, where we shall occupy the entire twenty-second floor, a penthouse, and part of another floor.

In Chicago, we have moved to the new building that we own on East Huron, near Michigan Boulevard.

As one can imagine, the frequent contacting of all these clients scattered all over the United States, in England, Germany, Holland, Belgium, France, Brazil, Mexico, etc., involves a large amount of traveling. We have our own transportation department, which takes efficient care of all reservations and sees to it that we get hotel rooms away from the traffic noise and bedrooms or compartments that are not over the wheels.

The publicity department, brilliantly directed by beautiful Betty Reese, is a very busy one. There are a great many requests from all over the world, literally, for information about our work or about certain products we design. The department also takes care of a comprehensive collection of slides and of all texts required by members of the firm who are making an address, an illustrated talk, or a complete lecture. There are several such talks every month, besides radio, television broadcasts, or newsreels.

In addition, technical papers or articles are constantly written for various publications in practically all fields of manufacturing, merchandising, or retailing, and in sales psychology.

John Grant, Office and Personnel Manager, is one of my closest friends and a delightful chap with a good sense of humor. He is also in charge of taking up with the various division heads the patents that are offered to us incessantly. Nearly 98 per cent of them are worthless, but we study them so we don't pass a good one.

He also answers letters which we receive from all over the world sent by slightly cracked people on the fringe. In general, they feel that their genius has been misunderstood by the world and that if we would only pay their transportation to the States (including the wife and six children) from Afghanistan, we could have all the rights to their new invention. The invention is usually something like a machine that can pick white hairs off camels and transplant them into black Himalayan rabbits in order to make fur sandals for the Patagonian market. We always answer, one way or another. Mostly one way.

Fellows with inventions are quite a pest, but as a rule we interview them. They are extremely tenacious and difficult to get rid of. In twenty years of industrial design practice, I haven't seen more than two or three patents of any value brought to us in that fashion. And they weren't worth much. The greatest majority of valuable patents are originated with established firms, or the research departments of large corporations.

We naturally take patents ourselves, an average of fifty a year, practically all of which are assigned to our clients.

Package design is a definite specialty in itself. If expertly done, it may be a factor of major importance in the success of a product or a concern. When Armour and Company, the great Chicago packing firm, asked us to redesign their packages, we made a preliminary study of the problem. It turned out that over eight hundred products had to be considered. The reduction to a simple design common denominator of several hundred disconnected labels, wrappers, boxes, drums, cans, cartons, bottles, and bags was a major undertaking. This study involved careful field investigation and research, which lasted nearly six months. When interviewed by a magazine writer doing a piece on the subject, Armour's Vice-President Walter S. Shafer said, "We didn't know what Loewy was doing." As a matter of fact, the result of this survey indicated the advisability of abandoning all the complicated and expensive multicolor labels in favor of a simpler, more striking two-color scheme. This potent design, carried throughout the company's products, saved enough money in color printing along to cancel out our designer's fee. Besides, it proved successful in the field, and we are constantly at work, improving further and adapting to more new products the original key design.

Food packaging reminds me of a charming, simple blond who complained to me that she did not get satisfactory results with frozen fried potatoes. On investigation, it turned out that she put the potatoes in a skillet to warm them up. But as she did not remove them from the cardboard package and the paraffin wrapper, the whole thing was a hopeless mess of cardboard and melted paraffin.

We should have mentioned on the label that results were best when the wrapper wasn't cooked together with the contents.

Now we see to it that instructions are printed clearly.

There are cases where industrial design has not been beneficial. Not long ago, the circus called on the profession for help. In spite of the designer's talent, the result was, in my opinion, one of the profession's authentic flops. A basic misconception of the circus queered up the whole thing at the start. The result was a pseudo-sophisticated mess halfway between stage and tanbark. All veddy, veddy distingué, all in pastel shades and monochromatic, all frightfully dull. One thing they couldn't Vogue up was the band.* Thank heaven it remained loud, corny, and continuous.

* Even this was attempted, and Stravinsky wrote a special score. However, it is now used in the concert halls only—not in the circus.

Which brings us to one of my pet resentments: the "sad clown." The "sad clown" situation is an acute one. I mentioned earlier the important place clowns occupied in my young boy's life. For a child with plenty of imagination, here was escape. They managed to transform matter-of-fact, earthy situations into dreamlike fantasy. A form of humor that was delightful and utterly unpredictable. But above all, they were fun. Their make-up was jolly, their routine hilarious. I remember having laughed myself sick, and grown-ups did too, watching their great slapstick. Their straightforward philosophy was that a clown should look funny, act funny, and *be* funny. Later, some sophisticated clown thought that it might be more refined to do funny things in a sad make-up. It is certainly true in the case of a genius like Charles Chaplin. But in the clown field, all I have seen to date is a bunch of desolate-looking artists going through ineptly sad routines. It is a matter of concern to me, as I believe that true clowns are important people, and that we should try to revive and protect their subtle slapstick, if I may use the expression. Men like the illustrious Emmet Kelly, that Knight of Gloom and Prince of Dejection, are just killing the art. Kelly is a puzzle to me. I have made it a point to attend the circus regularly in an effort to detect the reasons—if any—for his fame; this in spite of the fact that the new version of the circus bores me very much. I went through watching endless dozens of Armenian tumblers, Moroccan human pyramids, lady poodle-trainers in order to break through the Emmet Kelly fascination. The routine's nearest approach to humor seems to be the intermittent illumination of the sad character's bulbous nose.

I feel rather sorry for the poor sick kids in hospitals to whom

THE SAD CLOWN: The author was permitted to print this photograph "providing Mr. Kelly is identified as 'Emmett Kelly, world famous clown, under contract to David O. Selznick.' " Mr. Kelly's dejection seems to be of the contagious type. *D. C. Gunn-Graphic House, Inc. Photo.*

the great clown is supposed to bring cheer. If I were already sick, one good look at him would be enough to make me sink into the "critical list." I wish the talent of Kelly the clown could even remotely match the great heart of Kelly the man.

The offices of R. L. A. are not exactly orthodox as offices go. As most of our staff is made of artists and imaginative fellows, the general atmosphere is rather on the unusual side. Many of them have some personal interest besides design work, be it music, sculpture, females, cooking, camping, or Canasta. They collect things, write stuff, read fringe poetry and sometimes—in acute cases—write it themselves, but seldom inflict it on anyone. Many of them are excellent musicians, but they have too much creative impulse to accept using such commonplace sound-manufacturing equipment as a violin, a sax, a flute, or a piano. So they build their own and the results are weird. Once a year they play together, but they seldom agree on the tune, and the free-for-all concert would make good background music for a voodoo witch-splitting night in Haiti.

Practical jokes are constant and, occasionally, not in bad taste. The bulletin board, originally intended as a means of broadcasting interoffice intelligence, has become a display board for the latest jokes or cartoons. Whenever I pass by the board I am ready for a shock, as most of the cartoons are of the type that makes the management look a bit silly. My only possible retaliation is to find a cartoon that makes theirs look so tame that their sponsors feel like dopes.

A great many fellows like to work to music. So they keep a small radio at hand, playing all day long and—fearful of retaliation—in very low tones.

The management at Raymond Loewy's is particular about the appearance of its feminine collaborators, such as secretaries, receptionists, switchboard operators, etc. They are usually attractive and well dressed. My designers are themselves quite presentable fellows of education and taste, so the mixture is bound to be on the incandescent side. Interoffice romances leading to marriage are frequent and disruptive. But we love it. It keeps the place alive and gay. A favorable climate for a profession such as ours.

Another important factor in our operations are my yearly trips to Europe for business and pleasure. Our London office operates very successfully, thanks in part to the understanding attitude of Sir Stafford Cripps in reference to British exports. Our client list includes the largest corporations in the British Empire, including their subsidiaries in Canada, Africa, South Africa, Australia, Burma, Java, and Germany.

The chief value of these trips lies in the fact that they keep me in contact with European design development. I visit France, England, Switzerland, and Italy. There is a tremendous flow of new ideas generating in these countries, sometimes years ahead of America, and well worth watching. It is very refreshing and conducive to new approaches.

In 1938-39, I took a long trip through twelve Latin-American countries. Outside of some architectural ideas in Brazil and Uruguay, color schemes in Peru, the use of natural lava blocks in Mexico, and glassware in Buenos Aires, there isn't much for us to learn. However, the music is so exciting, the dances and the general rhythm of life are so different and unexpected, that the results are beneficial. This aesthetic shock treatment is conducive to the generation of new ideas, if not in function, at least in color, form, or texture. Mostly textures.

Once a year, shortly before Christmas, a great turbulence seizes the already sizzling offices of R. L. A. in New York. We start preparation for the annual show. This takes place in our own quarters, and it is usually very funny. A group of talented people, headed by Betty Reese, writes the sequences, the lyrics. Betty is a very good-looking young lady with the figure of a Tanagra and very, very witty. She appears in only one scene. Betty walks upstage sheathed in a tight-fitting white gown that accents her impeccable lines. Her blond hair is concealed under a closely tied white turban, her lips very rouged. She is attractive. Mr. William Snaith, my partner in charge of specialized architecture, then walks up to her carrying a large lemon meringue pie, which he proceeds to apply carefully and thoroughly against her face. Betty loves it.

Each year we invite a very few guests to our show. Among them are clients. They invariably enjoy it, and they understand our "let off steam" philosophy. It is normal and healthy, and makes everyone feel good afterwards.

We have frequent visitors at our office from all over America and from all over the world. About 1935 I met George Gershwin. We "meshed in" at once and it was delightful. He was then considering the erection of a studio in the Arizona desert, and we discussed it together. His conception was an unusual one, well adapted to the function, which was—naturally—musical composition. He wanted a large room of correct acoustics, entirely circular, with a slightly curved ceiling and no columns whatever. His idea of fenestration was most unusual: a thin slit, about ten inches in width, surrounding the whole room, approximately at eye level when seated. Properly glazed, this horizontal band would obliterate all foreground and most of the sky. In daytime, the view, over an unbroken arc of 360 degrees, would be a strip of distant horizon, where the desert would blend into the sky. Nothing puny about that conception, a true expression of a great man's mind.

Gershwin and I made countless sketches on restaurant menus and old envelopes. It was exciting and refreshing. He died before he had a chance to start anything on the project.

At the same time, Jean Cocteau, passing through New York, came to see me. As in the case of Gershwin, it was a memorable meeting. My office opened upon a large terrace handsomely landscaped with flowers, greenery, peach trees, and covered with white gravel. Comfortable wicker chairs were there and a low table. It was only two blocks from the then new Rockefeller Center skyscraper; the view was breath-taking and unobstructed. Cocteau liked to come at that exquisite moment of a fall evening when the buildings become illuminated in the golden twilight. We would sit and contemplate the magnificent sight far into the night—I, listening to the enchanting, dreamlike ramblings of a great poet in a great city.

In 1935, having assured a certain civil service gentleman that I was neither a bigamist nor a jailbird, and that I had no immediate plans to overthrow the government of the United States, my application for naturalization was accepted. In 1938, on a memorable day in my life, I became an American citizen. I had heard many an amusing story about things that happen during the long routine of naturalization. In my particular case, nothing happened that was even remotely funny. All I can remember is that on a sweltering day a couple of hundred sweating and nervous individuals were herded in a hot room, and that before we knew it we had become citizens of the greatest nation in the world. It was anticlimactic, and I regret it. I believe most of those present took their new citizenship as much to heart as I did, and we would have welcomed a bit of solemnity. Nothing stagy or tainted with pseudo-sentimentality, but some sort of short and impressive ceremony. We would have welcomed such a moment that one could remember and cherish forever.

Among an industrial designer's duties is the delivery of large quantities of lectures, preferably illustrated with slides, to all kinds of colleges, universities, professional organizations, Junior League meetings, business groups, etc. One of the questions most often asked is this: "Mr. Loewy, could you illustrate for us what is industrial design, really? Is it just styling? Or is it merchandising? Are you interested in the aesthetics of the problem?" Etc., etc.

I found out that a straight coverage of the question takes too long and is too technical. Conversely, the problem can be covered in a light vein and without pain. It isn't by any means the complete answer, but as a gag it carries its message pretty well. The starting point is usually any one of the famous Rube Goldberg's fantastic inventions. My point is that Rube is a great cartoonist but a poor industrial designer with no knowledge of the practical world. Let us, for instance, study the "Dishwasher" and the "Self-Sharpening Razor Blade."

Wind blows open door (A) pulling string (B) which causes hammer (C) to explode cap (D). Peaceful cockroach (E) loses balance from fright and falls into pail of water (F). Water splashes on washboard (G), soap (H) slides over surface, pulling string (I), yanking out prop from under shelf (J) and upsetting bowl (K). Goldfish (L) fall into bathtub (M) and hungry seagull (N) swoops down on them, thereby pulling string (O) which turns on switch (P) starting motor (Q) and causing razor blade (R) to move up and down along strap (S).

RAZOR BLADE

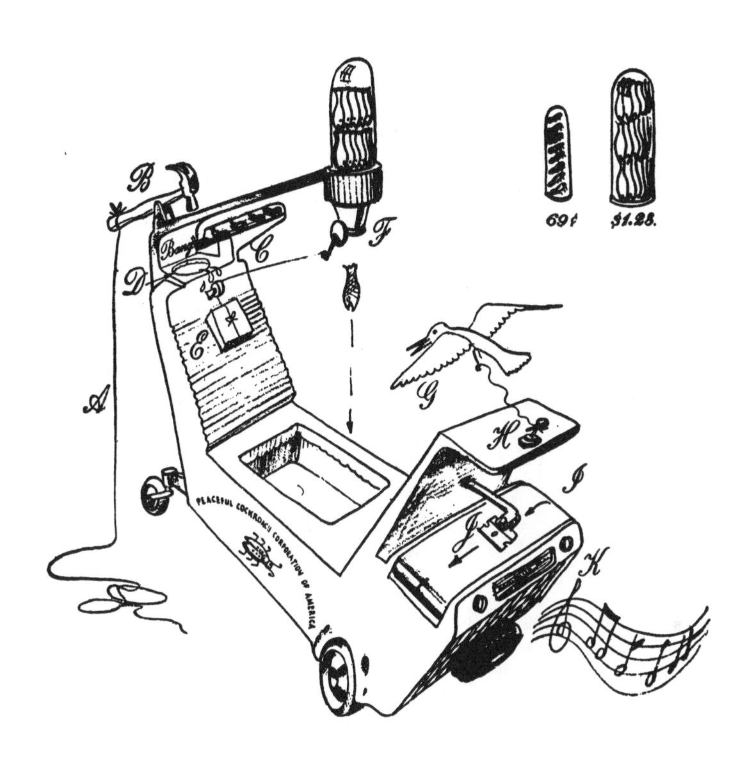

Mr. Goldberg's invention has been redesigned into one compact unit that can be purchased complete on easy installments. Cockroaches and goldfish are prepacked and retailed for 69¢ and $1.25 a refill. As an added feature a built-in record-playing attachment plays "On the Road to Mandalay" while the blade is being resharpened.*

(Made by the Peaceful Cockroach Co. of America) PAT. PEND.

* Balloon tires, $1.25 extra. De luxe model playing "Let Me Call You Sweet-Heart" and "O Sole Mio," $3.95 extra.

As in the previous machine, Mr. Goldberg has failed to take full advantage of its merchandising possibilities. In addition, his choice of the shivering agent (D) is not the best available and the opportunities for refill business are nonexistent.

DISH WASHING

This RLA improved model is completely self-contained and available F.O.B. Detroit at $99.50 in a choice of six colors. The shoe (B) is of the latest style. The shaking unit (E) is a Mexican Chihuahua with high-shaking coefficient. Instead of cold water this model drops automatically two dry ice pellets (D) inside an insulated plastic dog chamber. An interesting maintenance feature is added: After each use, motive power (dog) gets one aspirin tablet from seal pack (I) to prevent his catching cold. Refills are sold for 89¢ at all frozen food stores.*

(Made by the Dishpenser Company of the United States.)

* De luxe model with shoe spat, $1.00 extra; patent leather shoe for formal occasions, $2.95 extra.

CHAPTER NINETEEN

Already we have seen how attention can be attracted by a sudden change in the size or form of a recurrent design motif. (Chapter thirteen.) Another method is the change of pace and spacing as shown in this example. The result produces a feeling of deceleration and airiness. This technique is applied frequently in both product design and retail planning.

Chapter 19

CASE HISTORY

The reader may be interested in spending a day with me at my office. Let us choose a day in which we start on a design assignment for a new client. We shall have the opportunity to see an industrial designer in action. to follow up the design process and the development stage up to production time. You and I shall sit together, at my combination desk and drawing table. To justify your presence, you will be introduced to my client as Mr. Reader (or Miss Reader), my new secretary in charge of keeping minutes of the meeting. Our conference is at 11:00 A.M. The client: The Nadir Company of Boston, manufacturers of home-type ice-cream freezers. At 10:45, Miss Peters comes in.

"Mr. Loewy, Clare Hodgman and Herb Barnhart are waiting for a discussion of Nadir freezers."

"Ask them in."

R. L.: "Meet Mr. Reader, a friend of mine who will follow

the development of this product as a case study. Mr. Reader, meet Mr. Hodgman and Mr. Barnhart of our staff. As you know, Mr. Brown, President of Nadir, Mr. Grey, Chief Engineer, and Mr. Smith, Sales Manager, are coming for a preliminary discussion of their problem. Herb, did you get some material about the competition?"

Barnhart: "Yes. We have most of the leaflets, sales booklets, and literature. Also Sears and Ward catalogues. We have talked to salesmen in various stores. It seems that the Nadir is a good machine but it has not been selling well. The Iscold-Cream Freezer Company of Cincinnati put a new model on the market last fall, the Gem, and it has clicked with the public, at the expense of Nadir. Also, they say the Gem is less noisy, more compact, easier to keep clean."

Hodgman: "We have both the Nadir and the Gem downstairs. The other freezers are not important factors."

R. L.: "Have the Nadir and Gem sent up. When can you get started on this assignment?"

Hodgman: "Middle of next week. In the meantime, we'll get more information in the field, especially in small hardware stores."

Reader: "Can't the client give you that information?"

R. L.: "Yes, of course. And we shall ask him for it. However, we like to find out by ourselves too. Sometimes we have surprises. Even our clients admit it. Naturally, our own investigation is not an elaborate process; only a quick cross section as a double-check."

Barnhart: "I have suggested to the management that we study their wrappers, packing boxes, labels, and sales literature. Their leaflets are pretty bad and probably expensive. Also their displays at the point of sales. They are clumsy and rather old-fashioned."

R. L.: "Good. Miss Peters, let me see our contract with Nadir."

Peters: "It's in your folder."

R. L.: "Their stationery is very ordinary. And the trademark is just as bad."

Barnhart: "Yes, they need a potent trademark, badly. Besides, they use purple and yellow in their trademark."

R. L.: "Couldn't be a worse choice with women. I guess 90 per cent of the freezers are bought by women. Let's ask Smith during the meeting."

Hodgman: "The Nadir sells for $12.95. The new Gem for $12.50."

R. L.: "Herb, have a scale sent up so we can weigh them. Say, Clare, whom are you going to put on this?"

Hodgman: "Probably Jansen and Mercer. Jansen just finished the Unitex washer and he would be good on this one."

R. L.: "Okay, and keep me informed."

Our clients arrive. I introduce Clare Hodgman, in charge of the product design division. They already know Herb Barnhart, who will act as account executive.

The meeting soon discloses some background material. The Nadir Company is an old, reputable concern. They have built about the same machine for twenty-five years with some slight periodical improvements. The competition has come and gone. Nadir has stayed. However, a year ago Iscold-Cream put out the Gem, and it has been a sell-out from the start. Its success has kept up and it has hurt Nadir considerably. So we should start on the new program immediately.

R. L.: "I suggest that Hodgman, Jansen, one of our designers, and Mr. Reader visit your plant." They agree.

We make arrangements for the trip. Mr. Smith, the sales manager, points out various problems and tries to describe the machine in a leaflet. We produce his Nadir. Then he describes a certain feature of the Gem, trying to sketch it on a pad. We produce the Gem. He explains his point and we make notes. Mr. Brown asks me what I think of this Gem. I tell him and I add, "Another factor in its success might be that it is probably less noisy."

Brown: "Have you tried it?"

R. L.: "No, not yet."

Brown: "What made you say that?"

R. L.: "Salesmen told my men at Gimbel's and Bloomingdale's."

Grey: "Do you like its looks?"

R. L.: "It isn't bad but it is far from being right. Still looks too heavy. Let's weigh it."

We weigh both. Nadir is two and one-quarter pounds heavier than Gem.

R. L.: "I'd like to cut four pounds out of your present model. I believe it can be done. Also to make it easier to keep clean after usage."

Next step is for our task force to visit the plant. By this time, Mr. Reader, you know my boys well. You have exchanged, in the lounge car over some bourbon old-fashioneds, all the latest jokes around town, and you have shown each other snapshots of your wives, kids, etc. In Boston you have had a chance to meet other executives of the Nadir Company. All charming gentlemen who have taken you to lunch in the executives' cafeteria. During lunch (shrimp cocktail, creamed chicken with peas and carrots, ice cream, and coffee) you have exchanged more of the latest, or not so latest, jokes and mutually displayed with this new group snapshots of your respective kids, wives, etc. You have discussed football or baseball according to season and deplored in unison high taxes and the labor situation. Meanwhile, in New York, Chicago, Los Angeles, our men have interviewed the field and found out more things of interest.

When you return from Boston, our designers have a clear idea of our client's facilities and equipment. We know what they can produce and what is beyond their reach, and we will keep our designs within these manufacturing limitations. However, parallel to these designs we shall make some advance studies that might call for new machines or processes. They may offer advantages so marked that the extra equipment required to make them would prove to be a remunerative investment.

So we begin the first phase of our work. We now have the blueprints of the new freezer as developed by the engineering department and we start making rough sketch studies. When dozens of sketches are made, we hold a meeting with Hodgman, Jansen, and Barnhart and we still feel that our designs look unnecessarily bulky. If we could only stamp a certain unit instead of casting it we might reduce the bulk appreciably. Some new sketches are prepared along those lines and they look much better. Some call for structural modifications, so we consult by telephone or personal visit with the engineering department. It may be that different materials could be used and bring certain advantages. We make our suggestions to the Nadir people. Let's assume that they agree. We then decide among ourselves which of these preliminary studies seem most promising and we narrow our choice to three or four designs. When the ground seems reasonably clear, from a practical viewpoint, we start making clay models. These are prepared in our model shop, full size in the case of a small article such as a freezer.

After a week or so we have, say, three models completed in clay and ready for a first discussion with the Nadir people. We make an appointment and set a date for a visit to New York. (They usually wonder if we could get them seats for the latest musical comedy hit.)

This is the day of the meeting. This time our visitors are Mr. Grey, the Chief Engineer, and Mr. White, in charge of manufacturing (they have asked for three theater tickets; Mrs. Grey came along), and we start the discussion in our conference room.

The four clay models are on a display stand in the field of a battery of floodlights and spotlights that accentuate every detail of the design. Comfortable but not too deep chairs are arranged in front in a semicircle. Near each chair is an ashtray-stand.

Each model is discussed in turn, criticized to the limit, and notes are made. The meeting may be over in a few hours or it may last a couple of days. Finally, we agree on one or two models that will be further developed, and that will incorporate the latest suggestions. Our Nadir friends return to Boston.

Within a short time, the designs are corrected and we have another meeting among ourselves. If the designs look right—they usually do at this stage—we start making finished models. These are made of plaster, or wood, and metal and finished to the extreme. Metallic parts are chrome-plated, others are lacquered, the trademark is finely drawn—and the finished model looks exactly as if it were ready to be used. Then we take black and white photographs that show what kind of highlights and shadows we shall get in newspaper and magazine reproductions. We also take a series of color photographs.

If the results are satisfactory to us, we call in our client for a comprehensive presentation. The technique of this presentation is similar to the one at the showing of the first clay models. If any further modifications are agreed upon, we note them. Usually there are none at this stage. The next step is for Miss Peters to make a reservation for a good table at Copacabana. We seldom accompany our clients.

On the following day we prepare a set of accurate mechanical drawings which Hodgman and Jansen take to Boston together with the models. If everything is all right, they are left with the client, who starts making actual working models.

A few weeks later the Nadir Company calls a morning meeting at the plant. Our task force returns to Boston and we are given a demonstration of the three models in operation. It usually turns out that one is somewhat preferable to the others in some or several respects, cost being an important factor. Also noise, ease of cleaning, ease of operation, etc. If everyone, or nearly everyone, agrees that it is *the* design, the matter is settled and we go through the routine shrimp, chicken, and ice cream luncheon. If not, one or two of our men stay until corrections are made on the spot.

The Nadir Company is ready to start drawings and wood models which will eventually be sent to the tool and die makers. While this is under way, a number of finished hand-made models are being tested under every condition. They are dropped, banged, hit, pushed, pulled, knocked around, frozen to subzero temperatures, and heated to tropical conditions. In other words, used and abused to the extreme. If they can take it, the final drawings are submitted to us for double-check, then released, and dies are started.

Die-making may take a few weeks, as in this case, or up to a year as in the case of an automobile body.

When the first models are manufactured, we are called in again to verify everything, such as finish, color, trademark, etc. By that time we have designed display material, suggested a range of colors for the product, layouts for booklets and leaflets, etc. Often the client asks us to look at the advertising illustrations in order to make sure that the product is shown to best advantage.

Production is in full swing. Thousands of Nadirs, Model 51, are moving on the assembly line, packed and crated, and stored in warehouses until there are enough in stock to send at least a few to every dealer in the land.

A few days before this moment has arrived, meetings are held in every large city, in hotel ballrooms or convention halls. These meetings are more or less spectacular according to the importance of the company and of the product making its debut. In the case of the Nadir it is a medium-size affair attended by about a hundred dealers or distributors. Usually the press is invited and refreshments are served. There are plenty of refreshments.

In the case of a major product introduced by a leading corporation, guests may reach into the thousands and be treated to lunch. The menu usually consists of shrimp cocktail, creamed chicken with peas and carrots, ice cream, and coffee. There is plenty of talk about weather conditions, high taxes, competition, etc. Jokes, practical or impractical, are exchanged in enormous quantities.

After lunch, an executive of the company, or two or three, makes a rousing speech, the band plays "Hail, Hail, the Gang's All Here," "The Sidewalks of New York," and "Oh, What a Beautiful Morning." The chairman reads a telegram from the governor of the state. In extreme cases, "The Star-Spangled Banner" is executed.

By this time, the refreshments have taken effect and tobacco smoke reduces visibility to ceiling one. In a final blare of trumpets, the lights are turned out, spotlights hit the stage, the curtain rises, and the Nadir appears to the delirious audience in its smoke-veiled innocence. On its right is a gorgeous blonde in gold lamé evening gown, on the left a red-headed babe in Bikini suit, with exceptionally long, fluttering lashes. There is usually a moment of hushed silence, then a roar of appreciation at the beauty of the scene, mixed with many wolf calls and hiccups.

The curtain goes down in the regulation blaze of glory, and everybody scrambles out in order to be the first one to reach the checkroom, the gents' room, or the nearest bar. There is also a long queue in front of the phone booths in order to make urgent business calls or calls about tonight's business urge.

The waiters are already removing the debris and making room for the dog show that is to start at 7:00 P.M. sharp.

The Nadir, my child, has been born and is on its way to commercial glory—we hope.

The design of a passenger automobile represents industrial design at its very best. It calls for imagination, talent, and technical knowledge. In it we find practically all types of designing, from typical product designing (instrument board, hardware, hood ornaments), textile design (interiors), to lighting and acoustics, etc., etc. Good automobile designers make good industrial designers. Many of my best men started in the automobile field. They are an extraordinary lot; among all designers they are the ones with whom I have the most fun. As a group they form an incredible bunch of wild creatures who can't think of anything but automobile. They think of it morning, noon, and night. They think of it at breakfast, at lunch, at dinner, and—I fear—in bed. The married life of a real bebop car-hound's wife is anything but simple. One of my key men drove his young wife to such a nervous pitch by talking motorcars twenty-four hours a day that she left him five days after the wedding. I must admit that instead of the four-room house he had promised to buy as soon as they were married, he bought a one-room cottage with a four-car garage that used to be a repair shop. She returned to him after a while but the situation is precarious, to say the least. The seven thousand dollars he had saved went into the building of a terrific roadster that could do an easy 100 m.p.h. Unfortunately, it had no top and his wife resented getting drenched in a rainstorm, even at a 90 m.p.h. clip.

Once in a while when I am at Studebaker in South Bend, I take my crew to a little steak joint called Alby's where the T-bone sirloins and French fries are first-rate and whiskey sours are generous. The juke box operates nonstop while Phyllis, the boss's daughter, takes good care of us. The dinner is noisy and the conversation strictly automotive. The fried onions are unsurpassed. By the end of the meal, the tablecloth is covered with sketches of front fenders, roadsters, and bumpers, everybody is revved up, and there's high octane in the air.

The high spot of the P.M. is the ride back home. Everybody gets aboard his souped-up job and the race is on. Those forty blocks through South Bend suburbs make the Indianapolis race look like a wheel chair promenade at the old folks' home. With a dozen supercharged high-compression engines going full blast and cut-outs wide open, it makes one's hair stand on end, and I love it.

★

The boys have a nice sense of humor. When one of the experimental jobs I had built for my own use was completed, my chauffeur came to the plant to drive it back to New York. As usual, we added plenty of extra weight in the luggage compartment in order to break in the rear springs. Two boys packed in three hundred pounds of plaster bags, and came to tell me that the car was ready. "It is loaded," said one.

"He means it is plastered," said the other.

As soon as the war was over, two of my assistants and I flew to London in order to take steps for the reopening of our British branch, temporarily closed during the hostilities. We went to visit our friends and former clients in London, Birmingham, Coventry, etc. They were happy to see us and appreciated the fact that we were not forgetting them, even though we were very busy at home, owing to the furious postwar activity. My assistants and I were shocked at the conditions, at the unbelievable suffering that our British friends had had to endure for so long. They were half starved but nevertheless reluctant to mention what they went through. Their innate sense of good manners, however, made them realize that to overaccentuate restraint in talking of such trials might in itself verge on affectation. Gentlemen, they struck the correct balance with typical British dignity and some very fine humor.

Sir Stafford Cripps, formerly Chancellor of the Exchequer, was one with whom I discussed what our goals were in England. This was in 1946. I had no trouble in establishing the importance of correct appearance for British exports. He already felt just as I did, and we discussed most fields of manufacturing. Sir Stafford was especially interested in the automobile industry and we had a frank discussion of American versus British automotive philosophy. He promised his help during our period of reinstallation and he has kept his word.

A few days after I flew back to New York, I received a very peculiar letter, addressed longhand, in red ink, marked H.M.S. OFFICIAL. The envelope was very small. It was a reused envelope, that is, the name and address of the previous recipient had been crossed out so that it could be mailed again to someone else: me. It contained a charming note, written in the same flamboyant ink on a tiny sheet of paper, expressing thanks for my visit and the pleasure of having met me. It was signed Sir Stafford Cripps.

★

John Winant was then United States Ambassador to the United Kingdom. He expressed just as much interest as Sir Stafford in what we were trying to do. I was rather surprised to discover that he already knew a great deal about me and what Raymond Loewy Associates had been doing in England before the war. He volunteered his assistance and we discussed the postwar manufacturing world in general; he certainly knew a great deal on the subject. John Winant has often been compared, physically speaking, to Abraham Lincoln. I think it fair to say that the resemblance went even further; he was a great American, one whom I felt honored to have met . . . and, to some extent, interested in our goals.

★

The time, however, had not arrived for a resumption of our activities. There simply weren't enough basic materials to take care of both reconstruction and peacetime manufacturing. Every product made was reduced to bare essentials; this was the "austerity" era. So we decided to return home and wait awhile. We also waited awhile for air transportation . . . nine days! Nine days in which we had occasion to find out all about "austerity" food, "austerity" blues, and "austerity" cold feet.

Two years ago I was at the plant of the largest automobile manufacturers in England with one of my most talented and typically American automobile designers. His manners were correct in terms of Middle West dirt-racing but a bit surprising in the U. K. He was perfect in the pit of a racetrack or driving in the cockpit of a racer. Gifted with a gigantic dose of Chicago dialect, six feet tall, dressed a bit on the zoot-suit side, Tuck made quite a sensation while we had lunch with our dignified clients in the principals' (as they call it in England) dining room. Tuck was obsessed by the two butlers in tails whom he kept watching suspiciously throughout the lunch. Our hosts were delightful and entertained us with some subtle and rather hazy jokes to which we responded with polite but perplexed laughs. Tucker was just plain bushed, but he felt that he too should be entertaining his gracious hosts and embarked on a torrid and all too clear story about the red-headed cutie pie who lost her panties in the diner of the "Super-Chief."

After a moment of strained silence we returned to normal conversation, a little numb, but relieved.

Stratford-on-Avon is halfway between our client's works and London. The managing director asked us if we had ever seen the town. We hadn't. So he arranged for us to have a large saloon drive us to Stratford, where he had ordered by telephone a sumptuous dinner at the Shakespeare Hotel. Then we were to see a play acted in the Old Shakespeare Theatre by its famous stock company.

When we arrived at the lovely fifteenth-century hostelry, we were received as distinguished guests from America by the manager in cutaway, headwaiter in tails, and his assistant. They all spoke as if they were on stage acting a part in *Othello*, and we were impressed. Tuck hardly understood what they said and it became even more strained after we had had two or three pink gins. The test was when the magnificent butler came with the menu of our specially ordered dinner and he asked Tuck how he wanted his steak.

"Blue," said Tuck.

"I beg your pardon?"

"Blue, I said."

"I am so sorry, sir . . . ?"

"Make it bloody, I said, ya know what I mean, just kiss it on both cheeks."

When dinner was over, our Shakespearean butlers were both flustered and relieved. We drove to the theatre, ready for the real stuff. That night they were playing *Life With Father.*

And so to London, where, after two flats, our limping saloon arrived at 3:00 A.M. in a torrential rain. Tuck had insisted on having the fun of changing the tires himself, to the consternation of our gauntlet-gloved, Ascot-tied chauffeur.

In 1947 we felt that the time was at last right, and we resumed operations on Grosvenor Street, one block from the United States Embassy. Industrial design is very much on the minds of the British people, so much so that I have lately received several letters from the United Kingdom on which the post office cancellation mark bore the legend: "Good Design—Good Business." The Council of Industrial Design is government-sponsored and very active. I found, however, a major discrepancy between the amount of discussion on the subject of industrial design and the volume of work produced. Somehow, the subject has become a favorite topic for argumentation among the Chelsea Set. The agitation seems to be mostly as to "whether or not a designer is justified in giving the public what it wants even should the consumers' taste be short of all that could be wished for. Or should his professional integrity compel him to produce designs of highest aesthetic quality, even if the penalty were to be the failure and eventual disappearance of his client?" The opinion is divided and leads to some passionate after-dinner discussions that leave me (according to what I have eaten previously) in absolute rage or plain somnolence. I have pointed out to my ardent friends that, while they are gliding gracefully in the azure skies of the design stratosphere, we American designers just keep on plugging and giving the consumer what they want and plenty of it! Oh, naturally, naturally we endeavor to raise its aesthetic standard of taste, but we feel that this worthy goal rarely justifies placing our client in bankruptcy. We are all for better-looking products, consistent with better-looking sales curves and better employment.

This shockingly American philosophy has landed me in plenty of explosive situations, especially in a series of articles published in the London *Times* on the editorial page, where I was eventually and graciously allowed to defend the American viewpoint. Once during one of these academic discussions, the Chelsea boys asked what I thought of their theory of sales versus aesthetics. "If I may be allowed to quote your compatriot, Shakespeare," I said, "I would call it 'weary, stale, flat, and unprofitable.' " When they reminded me that the industrial designer should be ready, at all times, to sacrifice the interests of his client on the altar of pure aesthetics, I begged leave to quote Winston Churchill's "I have not become the King's First Minister in order to preside over the liquidation of the British Empire." "In my humble case," I said, "I have not become my client's design consultant in order to preside over the liquidation of his corporation."

All these academic discussions are bound to disappear progressively under the pressure of a very low-brow but nevertheless pertinent law—survival of the fittest!

Industrial design technique in England has not reached our degree of perfection as yet. A contributing factor may be the

BEFORE

INTERNATIONAL HARVESTER "FARMALL" TRACTOR

AFTER

BEFORE

INTERNATIONAL HARVESTER "TRACTRACTOR" CATERPILLAR

AFTER

(Designed in 1939)

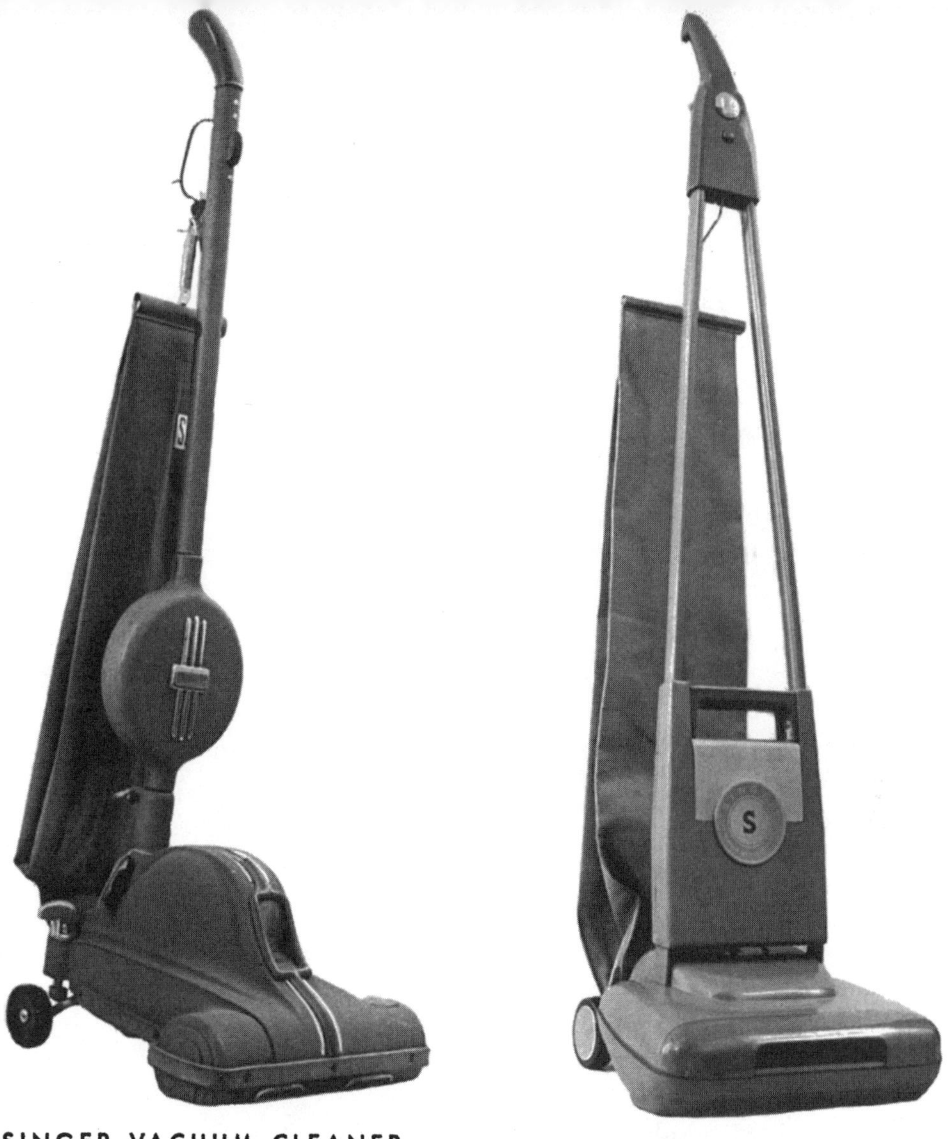

SINGER VACUUM CLEANER

When the unit on the left was brought to us, it looked somewhat as if Napoleon's hat had been pollinated by a horseshoe crab. Its appearance was not up to its excellent quality. The new cleaner, shown at the right, can be stored away easily. The cord reel has been resorbed. Its casing bears the trademark. Sales went up about 50 per cent and are increasing steadily. *Roy Stevens Photo.*

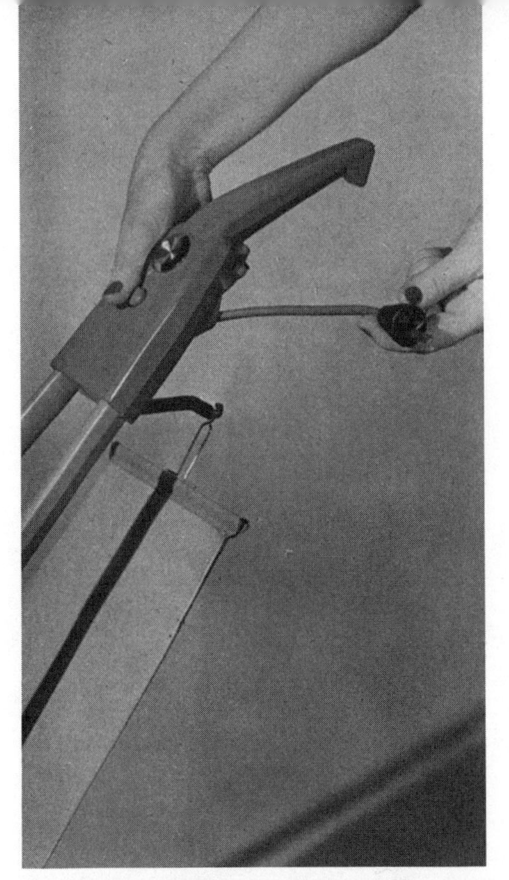

The cord slides through tubular shaft. The reel, which is now recessed between the dual tubes, contains more cord. *Howard Redell Photo.*

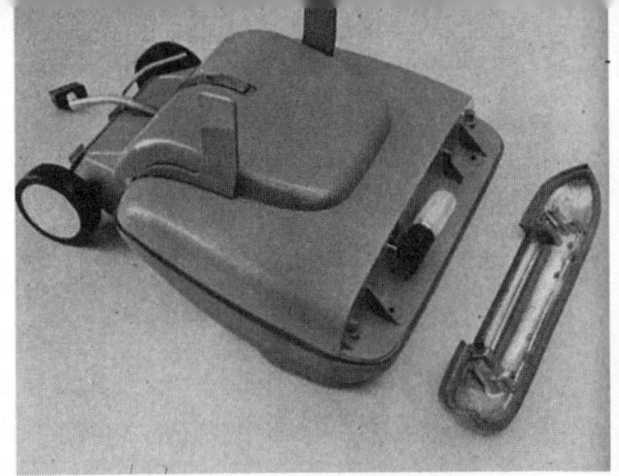

Top view of cleaning unit, indicating simplicity of design and assembly. *Roy Stevens Photo.*

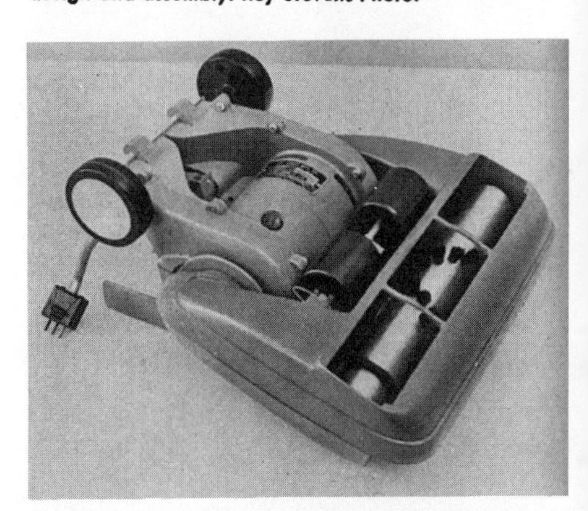

This view of the upturned cleaning unit indicates how the use of two small lateral fans instead of one large fan on top has cut down the total height. *Roy Stevens.*

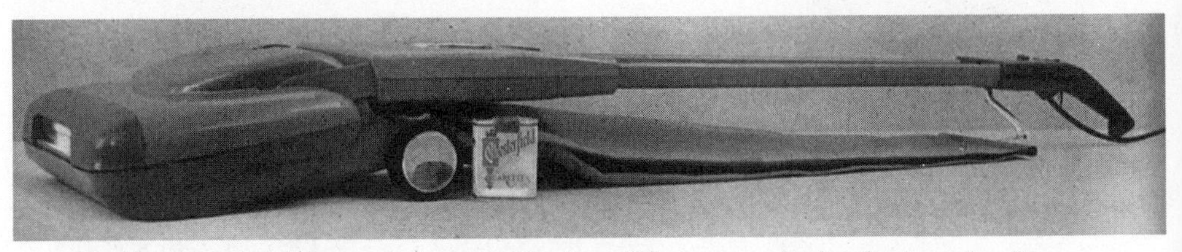

The cleaning unit is 50 per cent lower, making it possible to run under low furniture. *Roy Stevens.*

FRIGIDAIRE: The world's largest-selling household refrigerator. ***Robert E. Coates Photo.***

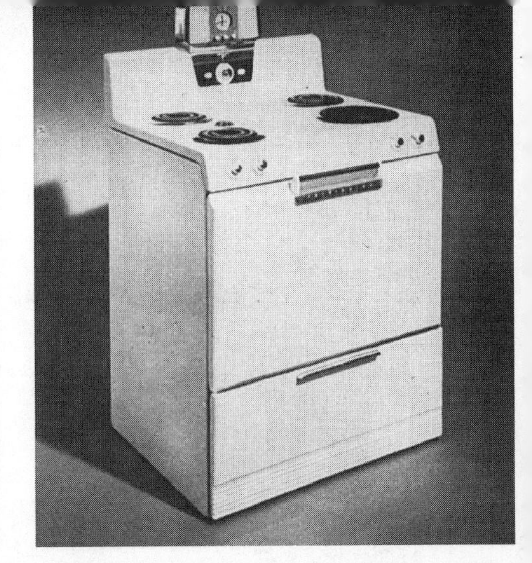

FRIGIDAIRE ELECTRIC RANGE: This one-oven range is the latest development in the field. Thirty-six inches wide, inexpensive, it is already a smash hit. Conceived by the research department of Frigidaire; Raymond Loewy Associates collaborated as designers and stylists. *Robert E. Coates Photo.*

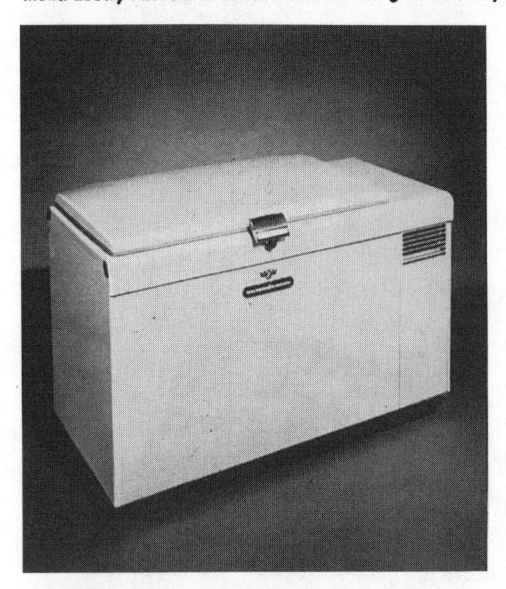

Frigidaire home freezer. *Robert E. Coates Photo.*

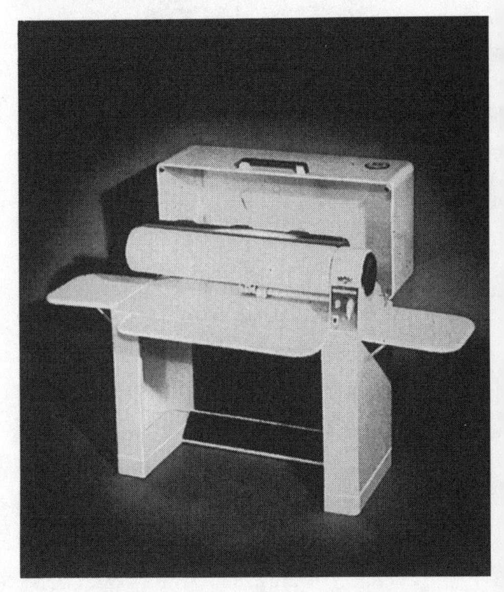

Frigidaire ironer. *Robert E. Coates Photo.*

AMERICAN CENTRAL KITCHEN CABINETS: Knee room at all work surfaces is provided by the extended top of base cabinets and sinks in the new American kitchens. The base of the cabinets is recessed for toe space. Rounded corners on all doors and drawers eliminate sharp edges. *Transfilm Photo.*

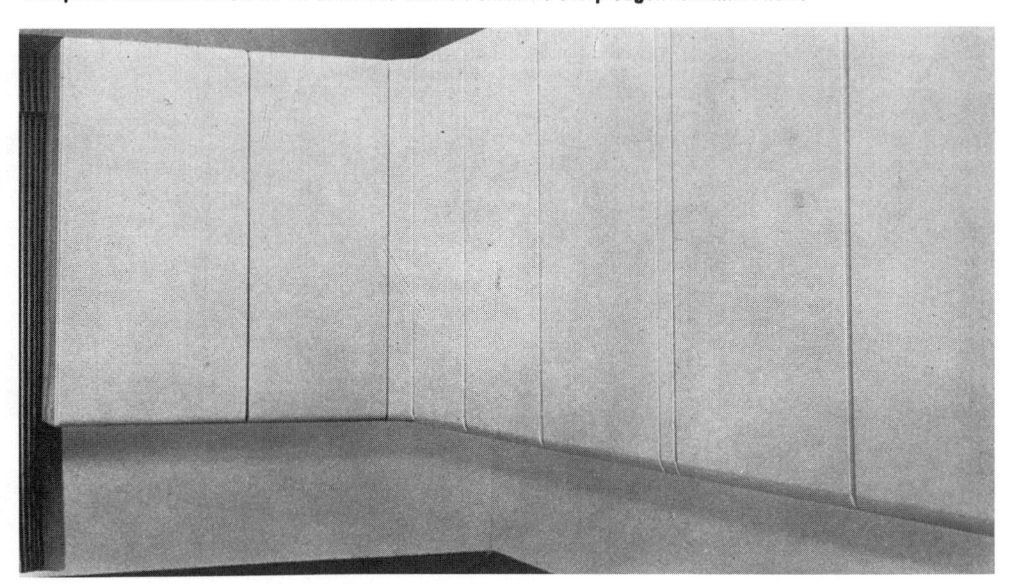

Concealed hinges, no hardware (fingertip pull) give simple appearance, cut manufacturing cost. *Transfilm.*

BEFORE **SOCONY-VACUUM** **AFTER**

HALLICRAFTERS

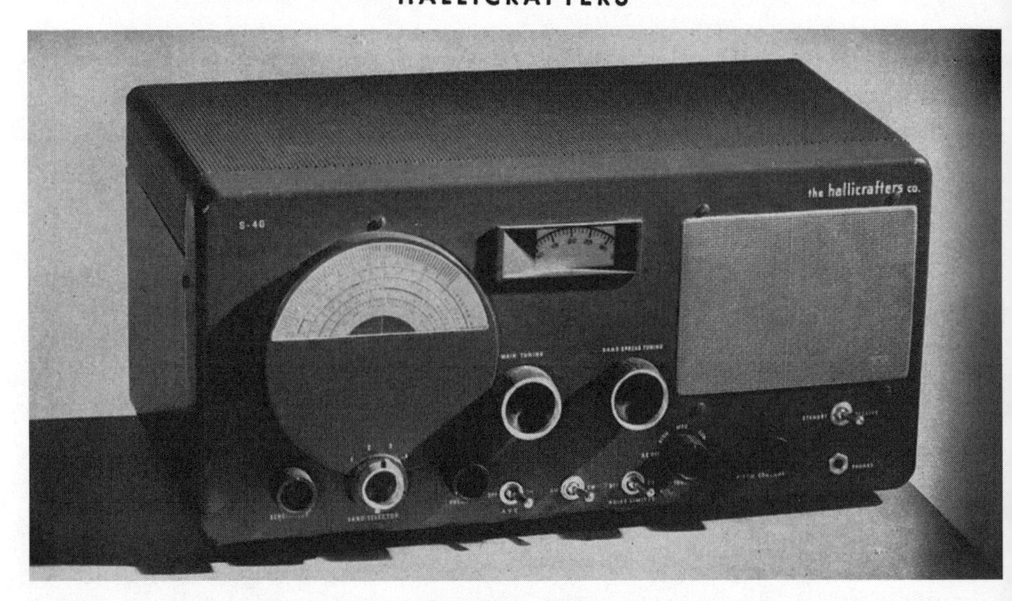

Short-wave radio set.

recentness of the profession; also the fact that credits are more limited and that there is still a certain lack of demand on the part of the manufacturers, store operators, etc.

But it is bound to improve quickly, as British designers have talent. Many of those working for our firm in London are excellent. What we, as employers, have to watch is their tendency to overstyling. When the design on which they work has reached a satisfactory stage, they seem unable to resist the urge to add one more strip, ornament, molding, twist, or curlicue; this spoils the whole design. In that they have to learn restraint from the Scandinavian, the French (lately), and especially the Italian. Some younger designers of the Milan school are outstanding in their straightforward design approach. Another weak spot about the British designers is that they are slow.

I am aware of the fact that some of my overseas friends may resent these mild criticisms. I hope they realize that they are meant to be constructive and nothing else. It is always a pleasure to work with British designers, whether those we employ or those we compete with: I know they will understand me and bear with me. As far as Raymond Loewy Associates is concerned, we are ready for the fullest co-operation.

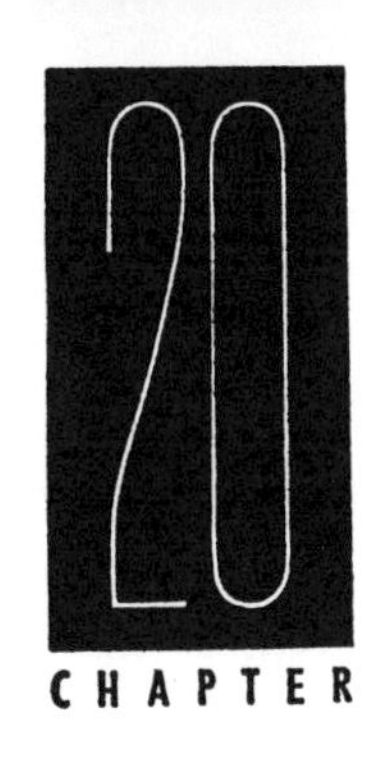

CONTRARY TO THE EXAMPLE ON THE TITLE PAGE OF CHAPTER 18, PROD-UCTS THAT HAVE MANY COMPOUND CURVES, SUCH FOR INSTANCE AS A TYPEWRITER, A SEWING MACH-INE, OR THE HOOD OF A TRUCK, RE-QUIRE SOME SEVERE ACCENT FOR RELIEF. THE WELCOME CONTRAST BETWEEN THIS FLOWING SCRIPT AND THE CHASTE BLACK PANEL ABOVE, ILLUSTRATES THE THEORY.

Chapter 20

THE MAYA STAGE

Being design consultants to one hundred and forty companies, most of them blue-ribbon corporations, and having been in very close contact with the consumer's reactions, we have been able to develop what I might call a fifth sense about public acceptance, whether it is the shape of a range, the layout of a store, the wrapper of a soap, the style of a car, or the color of a tugboat. This is the one phase of our profession that fascinates me no end. Our desire is naturally to give the buying public the most advanced product that research can develop and technology can produce. Unfortunately, it has been proved time and time again that such a product does not always sell well. There seems to be for each individual product (or service, or store, or package, etc.) a critical area at which the consumer's desire for novelty reaches what I might call the shock-zone. At that point the urge to buy reaches a plateau, and sometimes evolves into a resistance to buying. It is a sort of tug of war between attraction to the new and fear of the unfamiliar. The adult public's taste is not necessarily ready to accept the logical solutions to their requirements if this solution implies too vast a departure from what they have been conditioned into accepting as the norm. In other words, they will go only so far. Therefore, the smart industrial designer is the one who has a lucid understanding of where the shock-zone lies in each particular problem. At this point, a design has reached what I call the MAYA (Most Advanced Yet Acceptable) stage.

How far ahead can the designer go stylewise? This is the all-important question, the key to success or failure of a product. Its satisfactory solution calls for an understanding of the tastes of the American consumer.

We designers are realists; we like to deal in facts. Here, however, there are no yardsticks, no ways to chart a curve of public reaction to advanced design. Nevertheless, there are a few reasonably well-established facts among all these variables, and they can help us in our thinking. Being an engineer at heart, I have tried to introduce some order into this confused morass of human aesthetic behavior . . . in simple terms, design acceptance.

As a first groping effort in this direction I would like to express the Raymond Loewy Associates' ideas on the subject. One ought to bear in mind that the conclusions are necessarily empiric, and that while we speak of manufactured products in general, they apply more particularly to the automotive field.

1. Mass production of a successful given product by a powerful company over a period of time tends to establish the appearance of this particular item as the *norm* in its own field. (The public more or less accepts it as the standard for "looks" or styling.)

2. Any new design that departs abruptly from this norm involves a variable risk to its manufacturer. (We shall analyze the character of this risk later on; it has both positive and negative aspects.)
3. The risk increases as the square of the design gap between norm and advanced model in the case of a large manufacturer. (In plain language, for a big corporation a little style goes a long way.)
4. The risk increases as the cube of the design gap in the case of a smaller manufacturer or independent automobile manufacturer. (It is more difficult for these to establish a norm because they cannot blanket the nation with products in their design style.)
5. If the small manufacturer or the independent automobile maker succeeds in establishing a norm of his own, he may induce the large corporation to widen the design gap between their current and next models in an attempt to establish a new and different norm. Or, on the contrary, the large manufacturer may retaliate by reducing the gap in order to reaffirm forcefully the validity of his own norm and thereby discredit departing attempts of the competition. (Their salesmen will say to the prospective costumer, "I wouldn't buy that stuff, it is too extreme. You won't like it.") The big company usually carries the field through sheer weight of mass manufacture.
6. The consumer is influenced in his choice of styling by two opposing factors: (a) attraction to the new and (b) resistance to the unfamiliar. As Kettering said, "People are very open-minded about new things—so long as they are exactly like the old ones."

7. When resistance to the unfamiliar reaches the threshold of a shock-zone and resistance to buying sets in, the design in question has reached its MAYA stage—Most Advanced Yet Acceptable stage.
8. We might say that a product has reached the MAYA stage when 30 per cent (to pick an arbitrary figure) or more of the consumers express a negative reaction to acceptance.
9. If design seems too radical to the consumer, he resists it whether the design is a masterpiece or not. In other words, the intrinsic value of the design cannot overcome resistance to its radicality at the MAYA stage. There are some constants in the problem:
   a. The teen-age group is most receptive to advanced ideas.
   b. Two unmarried individuals each having high MAYA coefficients have a lower common coefficient as soon as they marry. (In other words, their collective taste becomes more orthodox, they fall into a bread-and-butter form of conservative buying habits.)
   c. The older age groups are influenced increasingly by the style opinion of the teen-age group. (This process is accelerating.)
   d. The wife is often the deciding factor at the time of the purchase. Her influence seems to decrease in direct proportion to the length of the marriage, reaches a plateau, and then reverses itself in later years.
   e. The MAYA stage varies according to topography, climate, season, level of income, etc. (For instance, an advanced design sells better in Texas than in North Dakota. Dark color is more popular in Pennsylvania than in Texas. A radical design will find good accept-

ance in larger cities, university towns, resorts; poor acceptance in mining towns, the farm belt, etc.)

In summary, let us say that any advanced design involves risk to the manufacturer. I believe there is no alternative between taking some degree of such risk or slow but certain eventual disappearance of the firm.

The smart manufacturer seems to be the one who is willing to take what General Eisenhower calls a "calculated risk." The theory expressed above, however empiric, may assist him in his calculation.

What does *risk* mean? There are as many definitions of it as there are persons. Our own thinking is somewhat as follows:

First, a large and successful corporation can get along year after year without taking more than a minimum of calculated risk. Its own norm acts as a style flywheel. This balance can be maintained until an equally heavy-weight corporation succeeds in establishing its norm along different lines.

Second, a smaller or independent manufacturer may survive for a considerable time on the basis of minimum risk so long as he follows the accepted style norm closely as it has been established by the leading manufacturer. But the company will not progress and forge ahead. A case of pernicious sales anemia sets in, resulting in eventual extermination.

Third, assuming that quality, engineering, and price are correct, calculated risk is the open gate to improved business for the smaller manufacturer. It is the key to successful operation and business expansion.

Fourth, the calculated risk should be such that it never takes the design beyond the MAYA stage, outside of some desperate business cases. These I call "Commando styling," borrowing the expression from the medical "Commando operation," where enormous sections of bone or tissue are removed from a condemned cancer patient as a last and desperate surgical gamble.

Fifth, there are reasonably accurate ways and means of pre-ascertaining the MAYA level of a given product in a given consumer's climate. This "climate" refers to: the state, location in the state, income, local characteristics, etc.

The above theory, tentative as it may be, has some value. We have made good use of it in the design of thousands of products, packages, structures, etc., for more than a hundred corporations.

Needless to say, the use of mathematical symbols—the cube, square root, etc.—does not pretend to be accurate. The terms are used figuratively to express a relationship which cannot be measured.

Guiding our clients in the virgin forest of advanced design is not our only task as trusted advisers. For instance, whenever a new product is developed and ready for the production okay, we use our imagination to the utmost in order to discover ahead of time any feature that might be distorted, misunderstood, or a subject for unpleasant jokes. To illustrate this point, we have stopped, at the last minute, a design for a household appliance that conveyed, under certain conditions of light, the semblance of a frog's head. Or a certain type of cooling unit, whose grille intake and two side knob controls mildly resembled a shark. Everyone remembers the case of a well-known automobile, an excellent car which, unfortunately, became known as the "Pregnant Six." This did not help sales. Or take the case of this particular book, which I first entitled "ENTER i." One reader might say that "all I got from ENTER i was a mild case of ENTER-i-TIS"; which would do the book no good.

As far as age groups are concerned, it is definitely established that the teen-age group is far more receptive to advanced design than the next age group: say from twenty to forty. After forty, the resistance increases fast. In other words, the dream of the alert industrial designer would be to design for teen-agers. But manufacturers realize that this is a low-revenue group, which is very regrettable.

I might venture to say that if the teen-age "design receptiveness" could be extended to the twenty-five-year age level, products could, in many cases, be manufactured along more advanced lines with assurance of easy sales. The older strata would soon be sold too, by contagion selling. But regardless of all this, I would say that the teen-age segment of America is of enormous benefit to the whole economy of the country in a great variety of fields and services, as it spurs advanced thinking. Its influence is many times superior to the limited buying power of the group. It acts as a booster to the whole country.

Personally, I have a deep affection for those boys and girls and I respect their taste. Whether or not they fall periodically for some silly fad that does harm to no one, their basic taste remains fundamentally correct. So more power to you, kids, the nerve center of the nation.

off-center treatments can be used

advantageously in the location of

controls, switches, knobs, levers,

etc., involving large areas. for

instance, in the case of a washing

machine or the instrument panel of

a passenger automobile, this com-

pact grouping is both aesthetically

correct and usually not expensive.

Chapter 21

THE BORAX PLAGUE

Do you like modernistic furniture? Does it please you? Can you live happily with it? Good. I am delighted for you and it satisfies me no end, as a modern designer and as a modern man. As far as I am concerned, it makes me sick and I'd rather live in a thatched-roof cottage in Cape Cod. There is nothing worse than bad modern. Bad modern represents about 90 per cent of the output, and it is tragic. Tragic and unfair to the multitude of young contemporary Americans who yearn to be up-to-date and to live with their times. Young married couples who consider past styles as dead styles and who courageously invest a great part of their savings in modernistic furniture. Poor misguided kids who discover a few years later that there is no style deader than bad modern. Then it is too late: they are stuck with the dreary stuff, and they must live with it. They have traded their precious savings for an arty mess of module-system junk. The module system is a wonderful idea of the bright *avant-garde* of contemporary designers. It is also called flexible furniture. Every piece, whether a commode, a bookcase, a bedside table, a buffet, a desk, etc., is of a standardized dimension, or module. So you can permute, arrange, and combine the units any way you like. People of imagination and physical restlessness can juggle the pieces around so that in infinite variations a poor setup can be transformed in a matter of minutes into other layouts just as poor. I have seen countless versions of the masterly, flexible furniture, and I must say that it all has a feeling of dread sterility.

It is a pity that there are so very few good pieces available, for we can't go on living forever with poor copies of Chippendale, Sheraton, or French provincial. What we need is a few inspired designers with far more than a system or a knowledge of modern materials and technology. The prime requisite is an understanding of charm, of the so-called amenities of life, a flair for the human instead of the dryly logical aspect of home surroundings. Some American designers have shown talent in this direction and several Scandinavian artists have grasped the idea. Unfortunately, the latter's Nordic creations do not seem to blend well into the stream of American life. They look best in Oslo, in Jönköping, or Malmö. In Philadelphia it is lost, in Little Rock it is poison.

Another pitfall into which some of the "modernistic" decorators have fallen with an acceleration of 3 G's is bad lighting. Being rational gentlemen, they have decreed that illumination, in order to be logical, must be evenly distributed, without highlights or shadows; in other words, it must be diffused. So we have

These ghastly things produced around 1920 served as inspiration for Borax Art. *Editions Eugene Moreau.*

BORAX DE LUXE. ***Wide World Photo.***

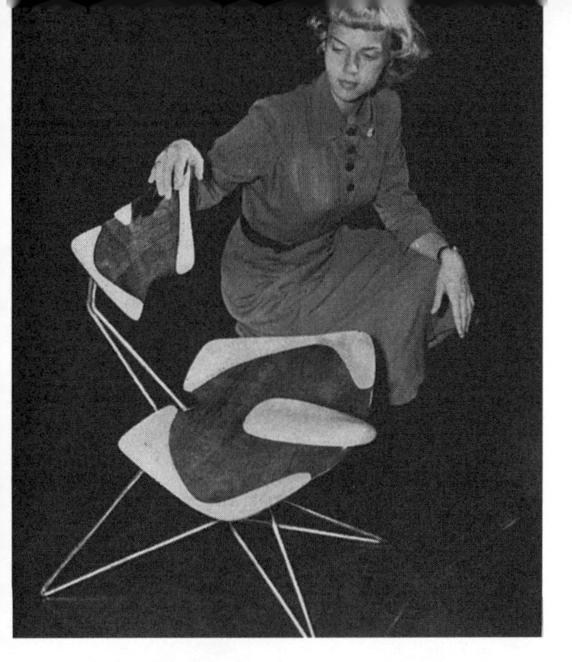

This mildly obscene "functional chair," like the one below, was designed in 1949. *Wide World.*

According to this machine's press release, it "enables the user to relax comfortably." *Wide World Photo.*

The antithesis of Borax. A well-proportioned object, designed with the same understanding of materials and stresses as an airplane wing or slender bridge. (Designed by Irving Sabo). ***Midori Photo.***

been flooded with that dismal invention called daylight fluorescent lighting. This frightful discovery produces a ghastly bluish halo that makes any healthy person look livid.* When used in a white kitchen with all white equipment, you'd expect to see Daddy stretched on the white enamel table ready for the postmortem. Junior seems to have caught a bad case of bubonic plague and Mother herself looks ready for the embalmer. The steak, when served, appears very, very gangrenous. Coffee is horrid and the mashed potatoes are a mass of blue-gray putrescence. It is all efficient, ghastly, and logical.

The living room is flooded with the same type of cosmic halo and the regimented module furniture stands at attention, all chrome knobs bright as buttons. Chairs are rigidly functional, and covered with some orange or poison-green fabric as a tribute to color. A low cocktail table with spindly legs and amoeba-shaped top introduces into the scheme the free-form element. On it are black ashtrays and some work of art—probably an elongated black panther. Or it may be a Nubian slave in gold turban ecstatically proffering another black ashtray.

Cubistic book-ends keep *Forever Amber* and *The Manatee* from falling over each other under the standardized stare of a Van Gogh in lithograph. (There is another Van Gogh across the room, the ineluctable sunflowers in a white frame.)

From a theoretical viewpoint the lighting is perfect: no shadows, no eyestrain, no dark areas. The furniture is efficient: all drawers, all shelves, all closets; no wasted space, no unnecessary projections. The heating is perfect, too. No wasteful fireplace, but a diffused, evenly distributed temperature. Nothing like a warm glow or a cool corner. Everything has been levelized at its logical

* Recent fluorescent tubes have a warmer tone and are much more acceptable.

level, reduced to its simplest technical expression. It is a wonderful machine to live in.

It is murder.

Before the war I had a top-flight Swedish designer by the name of Torben Müller. Torben had exquisite taste and he designed many lovely things for us. However, he would go, once in a while, on a functional binge that would lead to extraordinary results. He decided to have a truly utilitarian bedroom in his New York apartment. Reduced to essentials, a bedroom need not be large. It should comprise a bed, a chair, a door, and a window. Torben liked to read in bed. So he took a tiny room, the size of a large closet, and painted it all white in enamel paint. The floor was waxed and the window had frosted glass. In one corner was a hospital bed lacquered in white, crank and all. A dentist's floodlight projected a cone of bluish light at the exact spot where Torben would hold his reading matter. The only other piece of furniture was an operating room's tubular stool, also white-enameled.

I saw it once and felt depressed for days.

Poor, dear Torben, a delightful young chap, disappeared during the war while crossing the Pacific aboard a transport ship. One day he could not be found anywhere. Strangest of all, every single part of his clothing, personal effects, equipment, and weapons had vanished too, and no one had any idea of what had happened. A most functional disappearance.

To sum up this modern-furniture business, I think most of it is awful. I know this statement may get me into plenty of trouble. All I ask from my potential executioners is the favor of not being quoted out of context. *I am all for modern-design furniture* in theory. I own many delightful examples of it and my only regret is that there is so little that is any good. Bad modern is just as ugly as the revolting type of furniture that the trade calls Borax. For the reader unfamiliar with it, let us say that Borax is sold in enormous quantities throughout the country by cheap furniture stores. Heavy-handed in design, overstuffed, gaudy, and overstyled, it is the essence of furniture vulgarity. Usually upholstered with loud, sleazy fabrics, it is loaded with golden curlicues and polychromatic reliefs and sold to the lowest income group in staggering amounts. To the recently arrived Polish miner or the Croat foundry worker still obsessed by the misery of his Balkan background it represents materialistic splendor. They buy the stuff ravenously, take snapshots of the family sunken in upholstered American luxury, and mail them to Gdynia.

To me, bad modern is just as horrid as Borax, whether its perpetrators like it or not. The only difference is that in the first case the victim's name is John Smith instead of Borevitch Krzyck. Let's hope that someday the designer of modern furniture will think of it in a different vein, in terms of gracefulness and charm. Pieces that will help make the home a cozy place to live in, perhaps a bit less functional than a dissection lab or a dog hospital, but more inviting.

As to lighting, could we have it less engineeringly perfect? Could we retain pales of penumbra, realms of firelight and golden halo? Let's have contrasts, coziness, and warmth.

To realize fully the extent of the fluorescent catastrophe one has to travel through Latin America. I am thinking of a small village in Cuba that I have revisited after ten years. It used to be delightful, especially at night, when the colorful little corner bars, hardware stores, and fruit shops were aglow with blond light. It was sunny and gay. Every color retained its brilliance; it was an orgy of saffrons, apricots, mandarins, and magentas. It sparkled with joy. Now the whole village is standardized to the level of lividity, like other villages of Peru, Nicaragua, Chile, and Venezuela. The slide rule gents have done their damage. I doubt that their souls ever fluoresced in peace.

TOO GREAT A VARIATION OF DETAILS, WHETHER IN SIZE, FORM OR ARRANGEMENT, IS LIABLE TO CREATE UNREST AND A FEELING OF CONFUSION. THIS, EVEN THOUGH EACH COMPONENT ELEMENT IS AESTHETICALL

y correct in itself and by itself. Such is the case in this typographical example, where type faces are all good. The result, nevertheless, is

DISRUPTIVE AND IT BETRAYS LACK OF DISCIPLINE IN THE DESIGNER'S MIND. THIS IS QUITE DIFFERENT FROM LACK OF TASTE, WHICH INVARIABLY PRODUCES SCHMALTZ.

DESIGN AND PSYCHOLOGY

Now, dear reader, I would like you to be kind enough to look immediately at a certain page in this book and memorize *quickly* one—just one—of the numbers printed on it. The page is the next page. Please do it now before we proceed any further.

Go ahead, look.

1

2

3

4

If you have acted precisely as directed above, chances are that you have selected number three. The explanation of the mechanism of this trigger reaction is curious, but beyond the scope of this book. However, this example is only one of many that seem to indicate a startling possibility, the possibility of anticipating accurately the reaction of a normal individual to a given sequence of symbols. This theory can be used successfully by the alert industrial designer and incorporated into design, especially in the packaging field. I am constantly investigating the matter.

Little if anything has been written about psychology applied to design. This is one of the most fascinating aspects of the new profession, and one to which I often attract the attention of our designers. The sensory aspects of the normal human being should be taken into consideration in all forms of design. Let's take the perfect Coca-Cola bottle, for instance. Even when wet and cold, its twin-sphered body offers a delightful valley for the friendly hold of one's hand, a feel that is cozy and luscious. It is interesting to watch the almost caressing, affectionate way with which the average teen-ager fondles his coke bottle.

Or you may watch his daddy holding in his gently cupped hand the lovely globe of a snifter brandy glass. He warms it up lovingly, firmly pressing the stem against the sensitive inner part of his forked and outstretched middle fingers.

Chewing gum is another example. After a few minutes, the gooey mess loses practically all its flavor. Yet the addict keeps on chewing it for hours. This may act as a release from some sort of frustration. Under the masticator's deliberate and masterly will, the resisting body of chicle finally yields and flattens out in complete subservience, and this pleasant victory occurs at the rate of thirty defeats every minute. A defeat that is cruel, under the ineluctable pressure of the master's crushing molars. Each mastication is an assertion of victory over animated matter. One might almost say living matter, as chicle possesses malleability, warmth, moistness, resiliency, and intimacy of contact.

These, and many others, are examples of sensory designing of sorts. All happen to be very successful products.

A design characteristic that has become an R. L. A. trademark is "Tumblehome." Tumblehome shapes, which have been replacing the boxy, clumsy lines of the past, are nautical in origin. Long ago shipbuilders used the term to describe how a ship's sides, instead of being parallel, slope inward. They "tumble," or fall, "home." Tumblehome lines, if continued, would converge at an imaginary point, as in the case of a ship's sides. When you look up at a tall building and all lines appear to taper toward the sky, it is Tumblehome. In product designing we sometimes call it "accelerated perspective." When used with discretion, as in the case of an automobile body, a toaster, or the inside walls of a railroad car, it accents the feelings of grace and slenderness. We use it often, and successfully.

Few books have been written on industrial design, probably because of its recentness. Van Doren's *Industrial Design* and Teague's *Design This Day* are very different, but both good. However, there are many outstanding treatises on the related subject of technology and the modern world. Foremost among these is the brilliant *Technics and Civilization* by Lewis Mumford. Mr. Mumford covers the subject with a thoroughness that leaves very little for other authors to write about. To illustrate the wide scope of his vision on the subject of technics, may I mention a few of its more unusual expressions:

Amusement business, p. 315; Ax, p. 62; Balloon (Dirigible), p. 86; Balls and Syringes, p. 232; Bloodlust, p. 304; Cannon, p. 84; Contraception, p. 260; Conurbation, p. 163; Deflation, p. 400; Erotic (Life), p. 299; Ghosts, 195; Gnomes, p. 73; Fantasy (Abject), p. 300; Grundtwig (Bishop), p. 293; Qâsim (Abul), p. 22; Rape (Cold), p. 180; Sex (Tabus on), p. 260; Troy, Helen of, p. 245; Venus, p. 97., etc.

I usually recommend this book to prospective industrial designers in search of background material. It is most erudite and fascinating, and it can help them greatly. It is regrettable that such an outstanding treatise should become involved in political tirades about the dear old vested interests, the Wall Street vampires, etc., etc. The intelligent student will know how to skip this stuff and profit from the rest of the book.

Some shapes convey the impression of great density, viscosity, heaviness. Others express fluidity and lightness. Take a bottle, for instance. A slender, graceful bottle made of clear glass will connote lightness of body in whatever fluid it contains. A squatty, dark glass bottle will connote heaviness of body. To illustrate my point, here is an actual case history.

The X Brewing Company, one of the largest in the East, tested five hundred people at a party in Baltimore. There were all sorts of people—men, women, of all ages and professions. They were offered free an equal amount of slender, clear glass beer bottles and dark, squatty beer bottles. None was labeled. The testees were requested to state which bottle contained the lightest beer.

Ninety-eight per cent of the voters stated that the lightest beer was in the slender bottle. But it was a fact that both bottles contained the same beer.

Colors are also a major factor in design and they can be used to advantage. Some have a repellent effect, such as purples, violets, certain kinds of greens, and sulphur yellows. Others have a distinct effect on digestion. A scientist has measured the rate of food assimilation in thousands of guinea pigs fed the same menu in cages of different colors. It was found that certain pigments, especially in the red scale, practically stop digestion, while others accelerate the process.

Color can also make an object appear large or small. If two identical cubes are painted, one white, the other black, the white one will seem to be far larger. Colors may suggest safety or danger. The figures indicating miles per hour on some new automobile speedometers luminesce in shades from white to pink to bright red, as speed approaches the danger zone.

Colors connote levels of temperature. There are cool colors and hot ones. Manufacturers of plumbing fixtures who export their products to the far corners of the world identify the hot-water faucet with a fire-red spot, the cold one with a spot of iceberg-blue. We use this principle in the design of control knobs for refrigerators, electric ranges, and electric irons.

Colors of strong chroma, or brilliant colors, attract attention faster than dull ones. However, contrast plays an important role.

For instance, on a shelf packed with dozens of orange- and red-labeled cans, a white one would immediately stand out.

Various parts of the country show certain preferences. At the risk of being trite, let us say that cream-colored automobiles sell faster in Miami, Los Angeles, and Dallas than in Pittsburgh, Philadelphia, and Chicago.

When dealing in remote foreign countries, serious mistakes can occur and some manufacturers learn the hard way. For instance, there is the well-known case of a company who sent hundreds of thousands of white-handled hairbrushes to China. White is the color of death. The brushes were unsalable at any price.

In transportation, most design problems must take into consideration not alone physiological factors such as fatigue, tenseness, eyestrain, etc., but also the psychological ones such as confidence, stimulation, or annoyance.

Take the case of an aircraft interior, for instance. It is not sufficient to design comfortable chairs, good reading lights, and convenient footrests. It is just as important to select colors that will be soothing, decorative details of an unobtrusive nature or accents of an earthy, familiar nature that will favorably impress the inexperienced air traveler and thus improve his morale.

One might say that comfort is the result of correct physiological plus correct psychological design. More than an art, it is the psycho-physiological science. Some interesting research has been done in reference to the boundaries between physiological and psychological discomforts in air transportation. Awareness of these threshold areas is of prime importance in the designing of most vehicles: air-borne, steel-borne, water-borne, or rubber-borne.

As an illustration let us take the hypothetical case of two otherwise identical airliners whose interior accommodations have the same type of adjustable chairs, the same amount of light, an equal number of seats, similar baggage racks, bulkheads located at identical places, and an equivalent number of portholes.

In Airplane No. 1 chairs are upholstered in a glaucous green fabric. The window curtains right in front of your eyes are striped in green and red. The sides and roof are made of a herringbone pattern fabric. The floor is laid with a carpet of black and white checkerboards. Overhead, the baggage rack is a metallic net that shows its angular supporting brackets and bolts every eighteen inches. The reading light is only partially shielded and glares in certain directions. The end bulkhead is left bare, showing all its members and cross members, hundreds of rivets and bolts and joints.

The general effect, in physiological comfort, is entirely satisfactory. However, after four or five hours in such a vehicle the passenger would feel restless, annoyed, and generally unhappy. The psychological effect is disturbing.

Airplane No. 2's upholstery, walls, roof, and glass curtains are all in warm beige fabrics. The baggage rack is a long unbroken panel of perfectly smooth material, light beige in color. The window drapes are a light blue gray, and so is the plain over-all carpet. The reading light is well shielded and practically invisible at an angle. The end bulkhead is a smooth expanse of plain material in warm beige also. It is decorated with two simply framed color reproductions of good paintings. One represents rolling meadows under a lovely sky; the other one, a winding river among fruit trees in bloom.

It is evident that such an interior is not only as physiologically comfortable as that of Airplane No. 1, but it is also psychologically correct. It will inspire confidence; its familiar atmosphere will be conducive to relaxation and comfort; the net result will be a better morale.

We have seen earlier how psychology can be applied successfully in the field of retailing, as in the case of the "contagious buying philosophy."

Chapter 23

AUTOMOBILE BODY STYLING

"If you wish to converse with me, define your terms." VOLTAIRE

There is no phase of industrial design that is more satisfying and better integrated than the development of a new line of passenger automobiles. The Society of Automotive Engineers, a body of fifteen thousand technicians, is the foremost organization in the industry. Its standing and prestige are known and respected throughout the world. Once a year, the S. A. E. National Convention takes place in Detroit, and the Society has honored me with an invitation to be its guest speaker this year for the second time. The subject for my speech, as it was two years ago, is "Automobile Styling Today." For the reader who may be interested in the subject, I shall reproduce here, with the courtesy of the S. A. E., the text of this paper:

"Gentlemen:

"You have read, most of you, enough magazine articles about industrial designers so I don't have to describe this peculiar species of professional to you in detail. However, for those unfortunates who haven't experienced the joy, let us review the matter briefly.

"It is generally accepted that an industrial designer works best when draped at the edge of a turquoise swimming pool while Nubian slaves in gold sarongs serve chilled nectars in silver cups. Soft music relaxes his nerves while a blond, streamlined masseuse works on the master's right wrist. His wrist is a little stiff from having endorsed too many checks with a ball point pen. After being rubbed with pine oil, the industrial designer is ready for the morning chores, consisting of a series of interviews with the nation's foremost magazine writers in search of new material. He then makes ready for the afternoon session, which is devoted to press photographers.

"And so on and on and on.

"Since you are familiar with all this, let us concentrate on the sane moments when the master gets fed up with the routine and gets to the design part. For some strange reason my magazine writer friends are utterly uninterested in this terre-à-terre facet of the designer's life. What they want is the swimming pool story! So if you will bear with me we shall consider the master at work. The master, unromantically enough, gets up at seven A.M., eats little pig sausages, and starts plugging like any plain ordinary businessman.

"What he plugs at is design. Now there are designs and there is the imaginative, dream-type rendering, called blue-sky designing. These designs (generally made with air brush or pastels) are largely impractical fantasies of an exuberant artist. They have been severely criticized by practical engineers and others. Wild drawings are misleading to the public. As such they constitute a detriment to the profession of industrial design.

"However, I believe their effects are not so injurious as some engineers believe. There are a few 'crackpots' among designers, but our quota is not so different from that in other professional fields, such as medicine, law, architecture, etc. Furthermore, indiscriminate criticism is liable to stultify creative exploration. The complete elimination of imaginative designing would be a definite loss. Further, I doubt that it can be stopped. Unlimited imagination is a typical American trait. It can be channeled (to great advantage, as we have seen) into practical lines.

"So far as actual body design is concerned, we practical designers deplore just as much as the engineers the excessive bulk and weight of the modern passenger automobile. But instead of blaming the designer, why not single out our friends of the sales department? I know it sounds like first-class buck passing, but how many times have we heard the dear boys say to us, 'What the public wants is a big package,' and we know that a big package means a great deal of weight and a lot of money?

"Weight is the enemy. The average automobile weighs thirty-five hundred pounds. Thirty-five hundred pounds of materials to transport one or two people just does not make sense. Statistics show that ninety-two per cent of the cars on the highways travel with empty rear seats. The weight trend in the past years, I believe, has been decidedly retrogressive. This must change.

"So far as appearance is concerned, some designers favor rear engine propulsion because it eliminates the hood and brings the passengers forward. With this I do not agree. Hoods will remain as a protection to the forward passengers both for practical and psychological reasons. The hoods, however, will not be as bulky as before, and they will be designed to allow better visibility. This factor of improved visibility is vitally important when one considers that in the United States alone each year, forty thousand people are killed and two hundred thousand are hurt or crippled.

"What about appearance? Hysterical chrome-plated grilles are on the way out. They are costly, heavy, and ugly. Low intake scoops will replace them. Price of the car will probably not go down. I can't see aluminum or plastic bodies for quite some time—this, mostly from a cost viewpoint. Seats could be made much lighter, but here again we are up against cost.

"As far as the vehicle's general appearance is concerned, I believe that correct basic forms will produce better appearance automatically. Clean, definite highlights will take the place of applied chrome gadgets or body moldings. The trend moves toward the crisp, lean, racy thoroughbred instead of the bulky chrome-plated monster. Design will be fresh, clean, and young-looking."

What you have just read is an excerpt from the paper I delivered to the S. A. E. eight years ago, at the start of the war. It was written at a time when we were all pretty much in the dark as to what the future would bring designwise. I thought that you might be interested in this reminder that designers demonstrate some consistency of thinking.

Now let's consider the setup of an automobile styling division as of today.

The primary function of the styling division is to develop a design, in close collaboration with the engineering department, which will be as attractive as possible within the strict limitations of cost and other practical considerations. Design on this basis is down-to-earth; it guarantees definite results in a definite time. The secondary function of the stylist is to originate other designs constantly of a more advanced nature, free from inhibitions—uninfluenced by immediate criticism from whatever sources—in other words, the blue-sky stuff that I mentioned before. No one, least of all the serious stylist, proposes these experimentals as practical. However, they serve to inspire fresh thinking and to prevent falling into design ruts or grooves. They are intended to stimulate both engineers and designers—to startle them, if necessary, into entirely new channels of design-consciousness. Out of their subsequent work, many interesting developments are likely to emerge. In many instances these developments go into production. If the final form is considerably altered, it may still be traced to the essential elements of design in the much-abused blue-sky project. None of you needs to be reminded that many so-called ridiculous features of yesterday are accepted generally today.

In order to serve this dual function at the Studebaker plant, an organization is maintained under my direction. My responsibilities toward the management, as head of the styling division, are these: (1) to establish a long-range design philosophy for the company, (2) to develop a practical design for immediate use within the frame of that philosophy, and (3) to maintain a state of alertness and creativeness in the staff. In the case of the South Bend office, a resident manager, my assistant in charge, reports to me directly.

The styling division is made up of designers, modelers, and pattern makers. In order to assist these men in their work, we try to maintain a proper atmosphere or design climate. For instance, we like to have our designers working on several problems simultaneously. Besides the immediate problem—which is always the design of the next production job—we encourage the men to proceed with their advanced designs. By encouraging them to pursue dual programs a state of mental alertness is sustained that is reflected unconsciously in the design of the production job. Another contributing factor in maintaining the proper design climate is the relative freedom we allow in working hours. If the designer is held strictly to conventional business hours, somehow or other his work begins to resemble nine-to-five creation. He may hit on a good idea at just about closing time and, if he is in the mood, we don't restrain him from working part of the night. Obviously, it would be unfair to expect him to report at eight the next morning.

Third, we encourage these designers to get out of the plant, to travel occasionally. South Bend designers are brought to the New York office in rotation, where considerable creation activity occurs in widely divergent fields. By giving the men a chance to escape their regular surroundings, to work at a different pace, with different associates, their outlooks are refreshed. The world seems to bring fresh ideas and sustain their creative talents at peak level.

Physical accommodations in the body styling division must be planned carefully. Our present layout consists of four elements: offices, drafting room, modeling rooms, and pattern shop. It is desirable to segregate the plaster-modeling area, which unavoidably houses a messy operation. By the same token (and this is a job in itself!) we see to it that the clay-modeling area is kept clean and neat at all times. We like to have sufficient space for flexibility of arrangement. This way we can enlarge or condense any component unit quickly. When the full-size model or models are completed, they are usually shown to the management in a special display room equipped with a turntable and plenty of both diffused and condensed light.

A showing as a rule is a form of theatrical presentation with the spectators seated in rows of chairs before the revolving stage. When conditions permit, I prefer a method which I call the "flash display." In this method, a new model, full size, is concealed from the audience and unveiled suddenly in normal daylight at a distance of approximately 150 feet. In a vehicle design, I think that this first impression counts most dependably. Either the model clicks immediately or it doesn't. If the basic lines of the automobile are static, if the car looks "stopped," it means just one thing to me: the design is no good. No amount of chrome gadgets, trimmings, schmaltz, and spinach will give life to such a body design. This car is a dead pigeon. Conversely, the automobile that makes the best instant impression is one that looks alive as a leaping greyhound, charged with speed and motion even at rest. This car is a success.

After the first impression is made—*and it is imperative that the reaction be spontaneously enthusiastic*—the design staff has ample time to settle decorative details during standard showroom presentation. When my organization designed several makes of

British cars in England I used this method of flash display. I remember our design's being shown at Coventry on a playing field during a light rain. Despite the handicaps, the effect upon the management was most convincing in its down-to-earth integrity. It seems to me that too many showings take place at close range in a confined showroom. When this is allowed, attention is focused upon detail immediately, and on account of this diversion the general, basic feel for the *total* design is often sidetracked. I believe it is possible that one of the reasons so many cars of the past have looked heavy, bulky, and slow is that they were shown in this fashion initially. It always shocks me during a showing to see some of the men of the company walking right up to the machine to inspect a door handle, a hub cap, or a rear light. Fortunately, the two or three top executives with whom I am collaborating don't do that. They stay at a fair range and appraise the general look of the job.

Now back for a moment to a typical body styling problem in order to explain how we achieve the appearance that goes over at showings. The steps are orderly and consistent.

First, I brief the design team, trying to give a complete explanation and elucidation of the problem. As Hope once said, "Unless one is a genius, it is best to be intelligible." At this time the blueprint of the chassis we are to work upon (supplied by the engineering department) is studied. It specifies interior dimensions and clearances.

This briefing is very important. It is the stage at which I try to clarify absolutely the target at which we are aiming. This must be done in a lucid manner in order to insure that the design process will not degenerate into an erratic hit-or-miss process, wasteful of time and effort. In the latter case, the result betrays this hesitating approach, and the final designs are unconvincing; they lack integrity and that indefinable quality which means success. So let us say that the first requisite is that I myself, as responsible head of the styling division, should have a clear idea of what I want to achieve. As a famous French philosopher said two centuries ago, "What one conceives well one expresses clearly, and the words to say it appear easily."

What is, then, that design philosophy which underlies the whole project? In other words, what is the strategy that my tactical task force is directed to execute?

It is a simple one. I have believed for years that in the case of an independent manufacturer such as my client, Studebaker, reliance upon engineering, manufacturing, and business integrity alone is not sufficient for success. Stylewise I do not believe that keeping in step with the competition is enough. I believe that an independent, in order to succeed, must be courageous and progressive. The results may be somewhat of a shock, but it is far better than blandness. Blandness of design for an independent means pernicious sales anemia.

My staff designers are thoroughly indoctrinated in some fixed principles. The emphasis is always on the same points: (1) that weight must be kept to a minimum, (2) that visibility must be excellent, and (3) that the automobile must look fast and that, even stationary, the car must have built-in forward motion.

We add to these eternal criteria whatever new ideas may come from other sources—the management, the other manufacturers, public taste, and tendencies as they vary from year to year. There seems to be a certain cycle for taste in car design that we like to discuss. We are not too much affected by these tendencies, but they may be taken into consideration and reconciled to our stand-

STUDEBAKER EXPERIMENTAL CAR: The author frequently tries new ideas for cars, boats, houses, etc., by building a model for his own use.

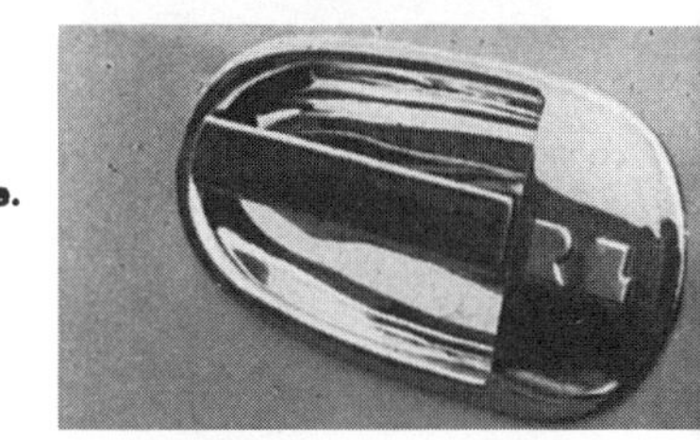

Detail of recessed door handle.
Roy Stevens Photo.

Features of this model have been adopted for production at a later date.

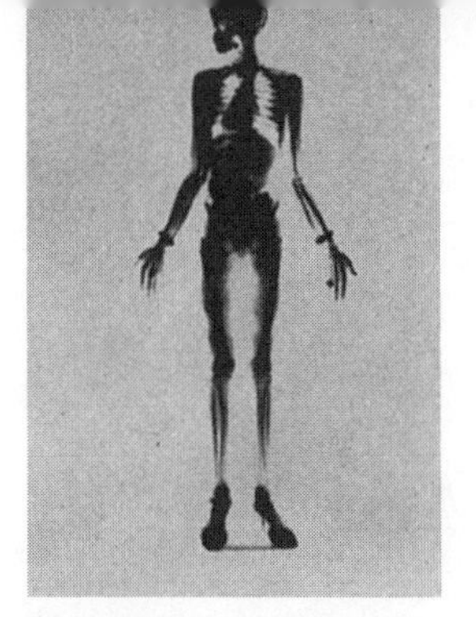

CHASSIS

AUTOMOBILE BODY DESIGN, based upon a chassis (or skeleton), obeys the same aesthetic canons of slenderness and economy of means as the human figure. There is such a thing as over-styling and bulbiferousness. Even when you look at these illustrations upside down or sidewise, the results are equally pleasing or repellent.

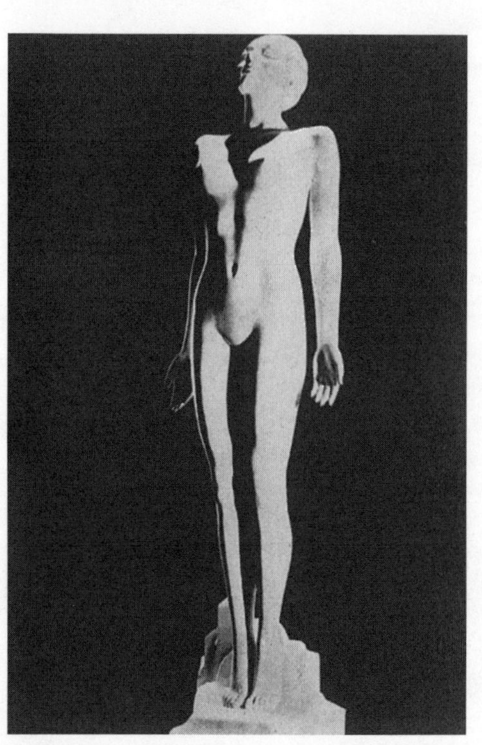

BODY A

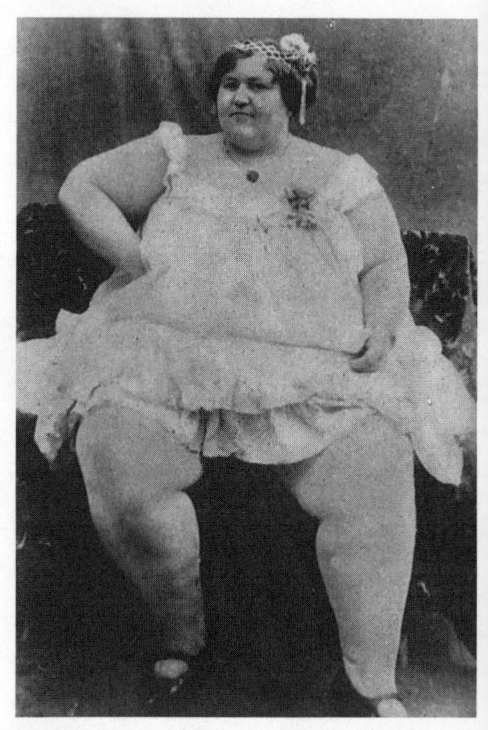

BODY B

CHASSIS

BODY A

BODY B

I. First rough sketch

II. Air brush rendering

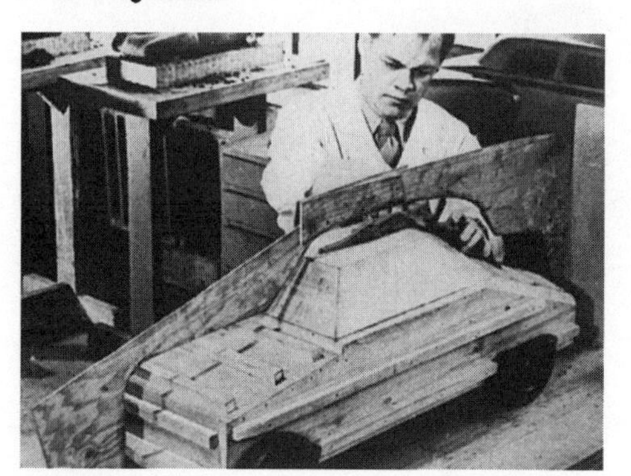

III. Start of the quarter-size clay model

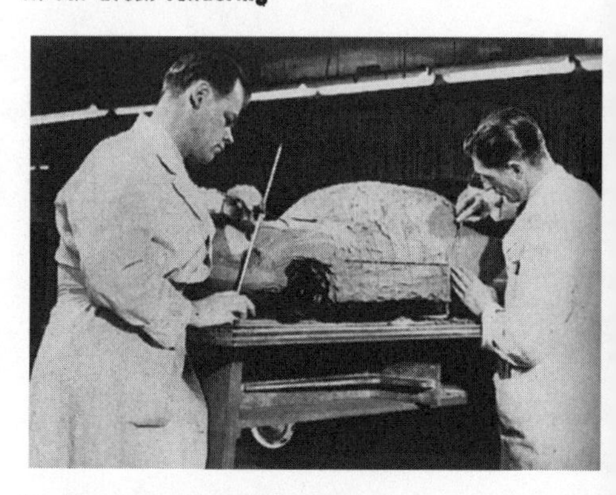

IV. The clay model is being worked out

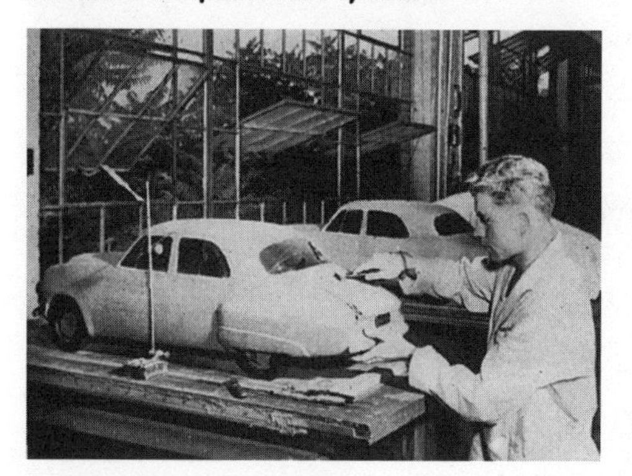

V. Clay model nearly completed

VI. Plaster cast is made from clay model

VII. Plaster cast is painted

VIII. Finished plaster models

IX. Full-size wood mockup is started

X. Then tested for interior dimensions

XI. Full-size clay mockup

XII. Full-size plaster mockup

''JELLY MOLD'': If not for the denture we might apply Shakespeare's ''sans teeth, sans eyes, sans taste, sans everything'' to this composite of the American ''fat car.'' Europeans call this frontal appearance ''the dollar grin.''

One example, among the hundreds of rough sketch studies, made at the inception of a new automotive design problem and usually designated "an automobile of the future."

THE FINISHED PRODUCT

ards of appearance if it is deemed advisable. May I mention in passing that I often discuss all these matters with men in the department who are not designers, with the modelers and pattern shop men. These fellows have more than the usual amount of taste; they are an important segment of our design team.

Next the tasks are assigned. A group of designers may be given the task of doing rough sketches of the entire car. Another group will work on front and rear ends only, etc. We establish a time schedule, deadlines, and the work begins. Within a reasonable time we have piles of rough sketches . . . enough ideas for half a dozen companies . . . and with the sketches several one-eighth-size rough clay models. Some designers prefer to work in clay, others with pencil. We impose no restrictions.

Now the *important* process of elimination begins. From the roughs, I select the designs that indicate germinal direction. Those that show the greatest promise are studied in detail, and these in turn are used in combination or arrangements with one another. A promising front treatment can be tried in combination with a likely side elevation sketch, etc. From this a new set of designs emerges. These are then sketched in detail. After careful analysis, they boil down to four or five. Of these, one-quarter-scale clay models are started. Again I want to remind you, these same designers are working on any advanced idea which interests them, and they may be working out this blue-sky stuff in one-eighth-scale clay roughs or in sketch form.

When models are satisfactory we cast them in plaster, paint them, and trim them with bumpers, door handles, lucite headlights, etc. They look like enticing, small, complete automobiles. They are all practical and within clearance and manufacturing limitations.

Now we are ready to select the best designs for full-size mock-ups. When it is decided what model to build, a full-scale automobile is made in clay, a method which permits correction of the distortions that invariably occur in the process of "scaling up" from one-quarter-inch size. After these corrections are made, a full-size plaster cast or a wood mock-up is made. This is then painted, trimmed, equipped, and glazed, and the completed model is an exact reproduction of the eventual production job. We find that a full-size plaster model has advantages over the wood mock-up in that it can be altered more quickly and economically. Compensating for this fact, though, it is true that a plaster model cannot be trimmed inside, and doors cannot open as they can on the wooden model.

I have found that when several models are to be shown, it is advisable to paint them all the same color so that color preferences will not influence the choice of the management unduly.

Generally it is desirable to have competitive cars available in or near the showroom for reference purposes. These will have been painted in the same color as the mock-up and will be shown in the same light at the same distance.

Now comes the time for the management to make the decision. Changes are inevitably suggested and another complete showing is arranged to demonstrate how these alterations have been incorporated in the design. When the final okay for production is given the design cycle is complete. It is up to the engineering and production departments to draft it and detail it.

Let us review design trends; what they are, what they seem to indicate for the future. Let's consider the front-end treatments.

First, we have the Great Chrome Fence school of front ends. It all started with a little wafflelike grille, then it extended up, down, and sideways until we reach the Monumental Chrome-Plated Barrier. At this point, the fence is doing a good job of protection, not alone of the hood and fenders, but of a large part of the neighborhood. The job fencing being completed, we have reached a limit; and we may expect the trend to reverse itself and end up where it started, right at the waffle stage.

As far as we stylists are concerned, these results are negative, as they do not indicate anything new.

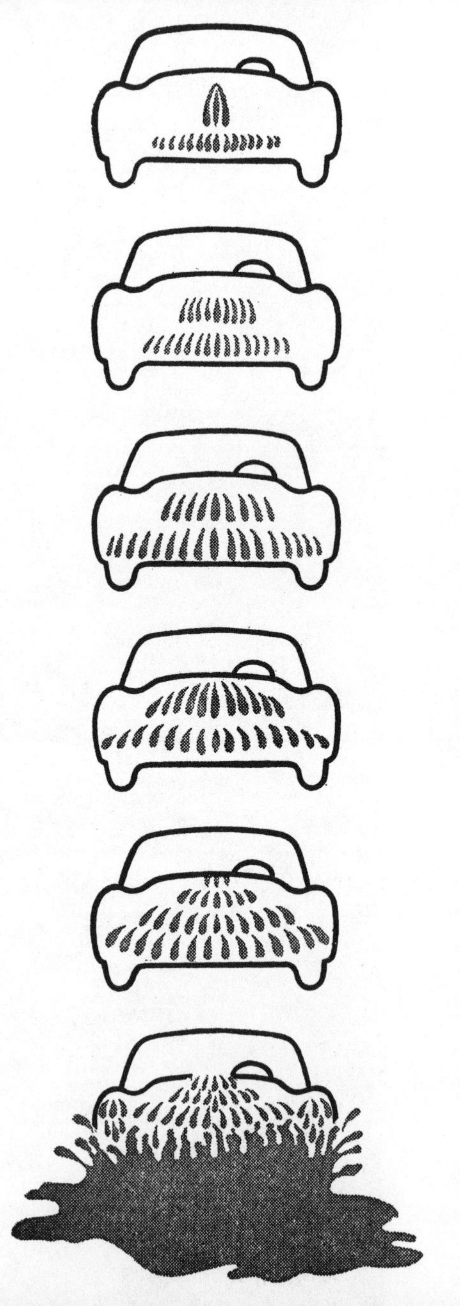

Then there is the Great Chrome Cataract school. It seems to have started with a tiny chrome leak on top of the hood. Unchecked, it became more and more leaky, spreading out right and left and creating secondary cascades. It ends up in a grand finale, the Great Chrome-Plated Niagara. Here too we seem to have reached a climax. As in the old-fashioned movies that were run in reverse, showing the swimmer diving out of the pool and landing up on the springboard, we may watch the sudden reabsorption of the chrome cataract, sucked up from the ground and back to a slow leak.

For the stylist, this study is still negative, and of no help whatsoever.

Now let us have a look at what the foreign competition has to offer. In every case, we are going to consider a reasonably accurate composite picture of typical foreign automobiles.

FRANCE

DELAHOARE

Boudoir Supreme

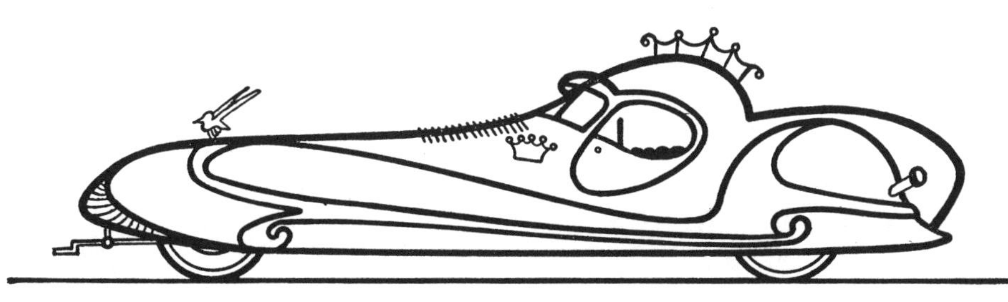

Very low, very long, de luxe and sensuous. Perhaps a little loud for our taste. Color: cream. Hot and cold running water.

ITALY

CHEESITALIA

Turboramoter

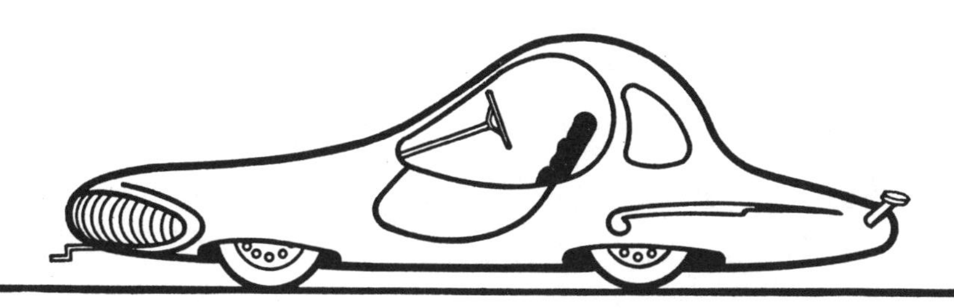

Extremely low, functional, and cramped. Color: black. Three speeds in reverse for quick parking.

TCHECKOSLOVAKIA

GRDNYATZCK FOUR

Beetle Scoopster

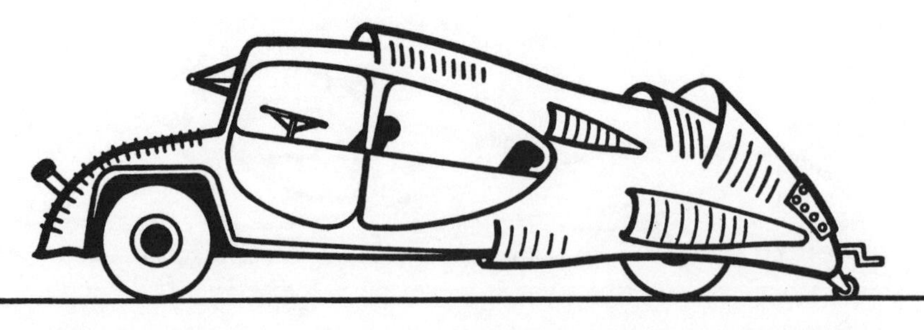

A super functional rear engine automobile. Very rational and very hard to sell.

UNITED KINGDOM

PANTHER

"Blue Tit" Saloon

(FOUR DOOR TUDOR)

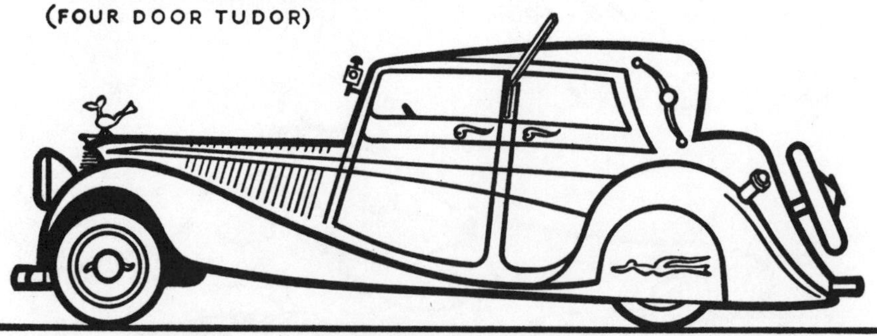

Following the British fashion of naming cars after birds (the Swallow, the Snipe, the Swan, the Albatross, the Falcon, etc.), the "Blue Tit" shows little that is new except its name.

DRAKCAP 72

The Soviet Union's products such as the DRAKCAP are strongly reminiscent of the PACKARD 27 (see below).

27 PACKARD

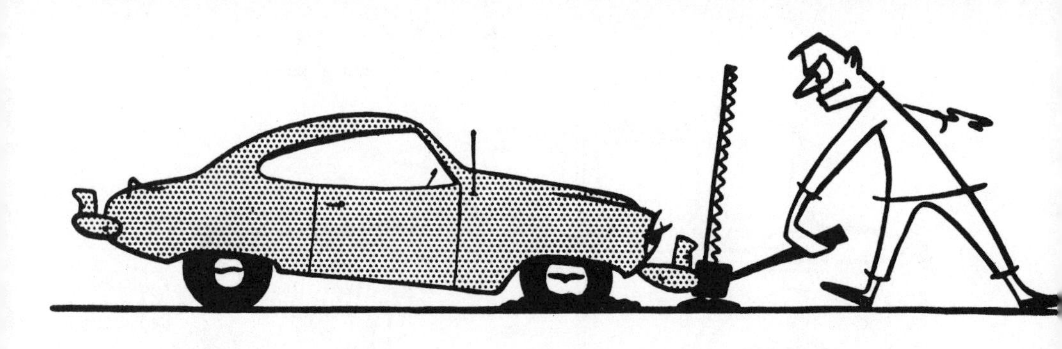

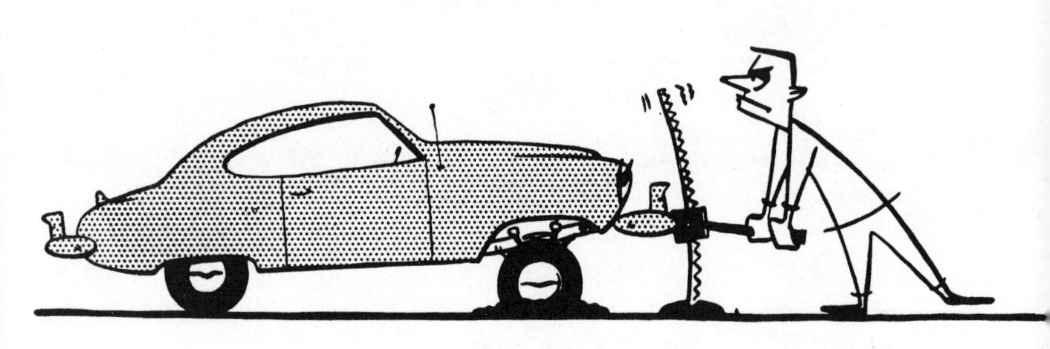

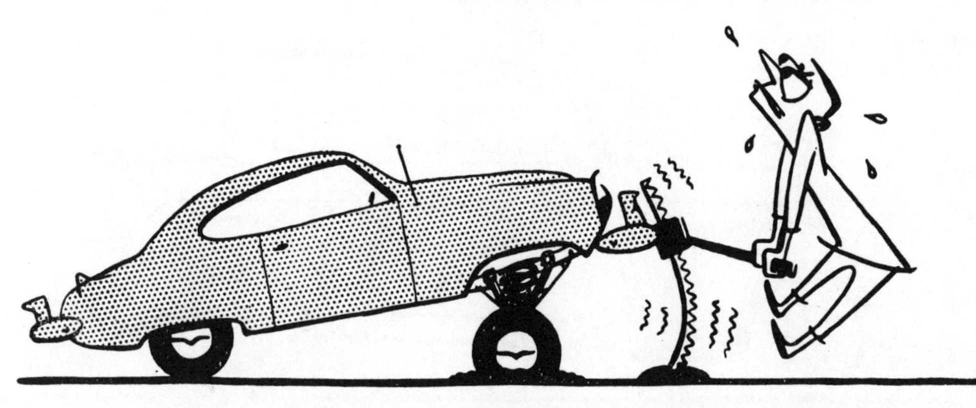

STRANGE BEHAVIOR OF THE JACKED-UP AUTOMOBILE

OR

THE CASE OF THE RELUCTANT FLAT.

SS *Panama,* prototype for three sister ships, designed in 1937.

Main Lounge, SS *Ancon.*

Standard first-class cabin set up for daytime use. ***Matson Lines Photo.***

Nighttime setup. Shower on right, washbasin and w.c. on left. Ship is air-conditioned. *Matson Lines Photo.*

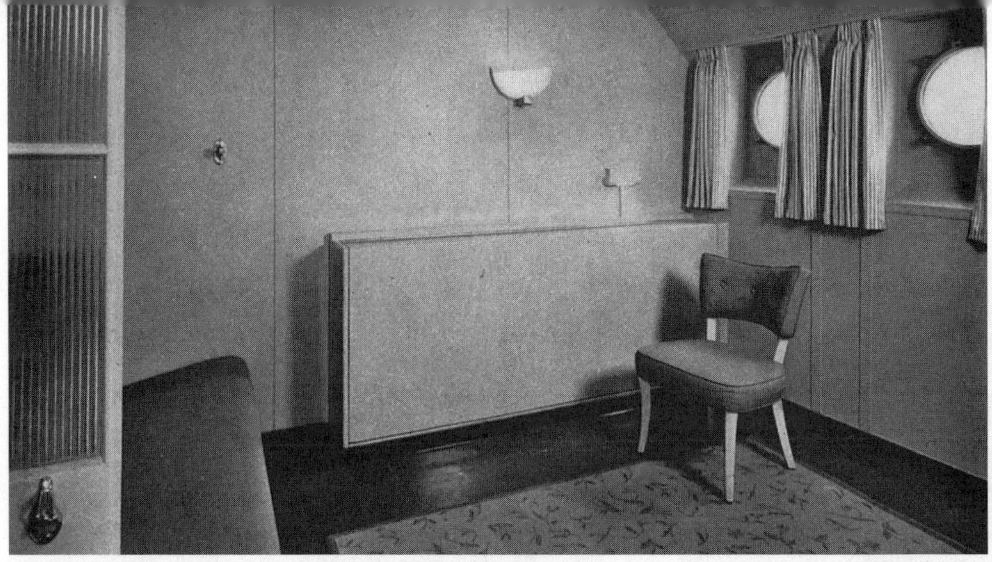

Cabin-class stateroom. Day and night setups. ***Matson Lines Photo.***

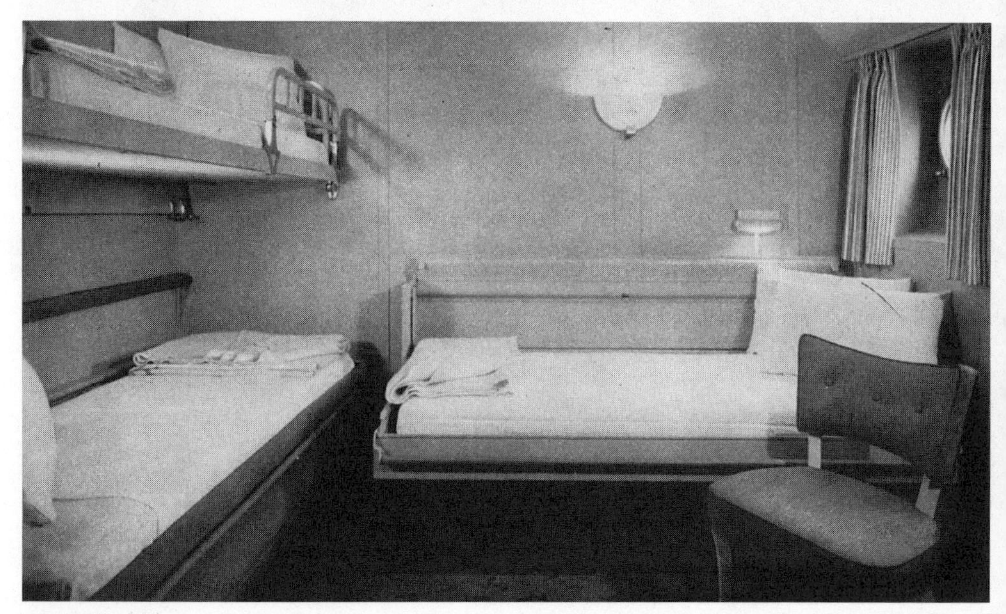

Matson Lines Photo.

MODEL OF A TRANSATLANTIC LINER: Exhaust from the power plant is through the two high, slanted stems on port and starboard sides of the hull. Smoke is thereby kept off the recreation deck.

Center superstructure forward replaces conventional funnel, is used as a climatized winter garden. Deck areas and location of lounges, dining rooms, etc., make this ship convertible into auxiliary aircraft carrier service.

The same process used in the design of a passenger automobile is adopted in the case of a locomotive . . . in this instance, the Pennsylvania Railroad S-1 prototype engine. This 6,000 HP locomotive is capable of pulling sixteen Pullman cars at a sustained speed of 100 m.p.h. with a top speed of 120 m.p.h. Designed in 1937 with the co-operation of the Pennsylvania Railroad engineering staff.

A series of rough sketches in pencil establishes a general appearance trend. Among these, the most interesting one (left) is retained for further study. This sketch is approximately twelve inches wide.

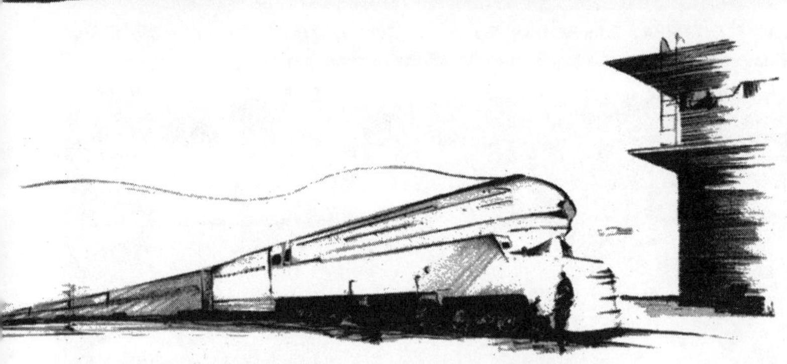

Another sketch shows the neat tie-up between engine, tender and consist (a term that means the different type of cars of which a train consists).

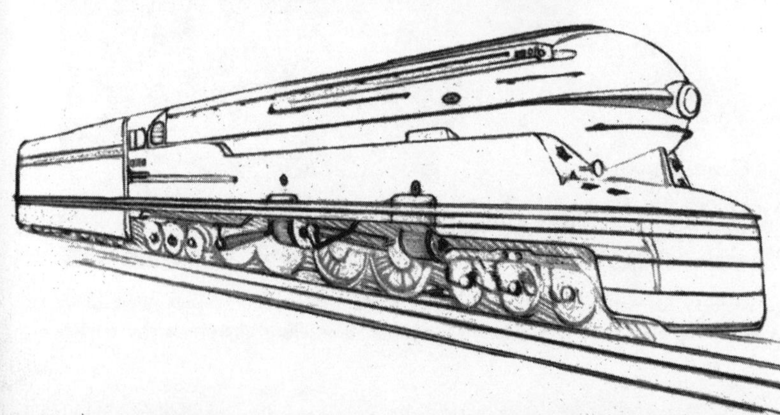

A more detailed sketch indicates the form to be taken by the major component units of the locomotive in relation to the whole

A clay model is made based upon the dimensions and other characteristics supplied by the Engineering Department of the railroad.

After the model has been checked and okayed by the railroad, we prepare for them a mechanical design showing both front and side elevations.

The completed locomotive, 140 feet long, shows very close design relationship to the blueprint, mockup and even to the preliminary sketches.

Raymond Loewy aboard the 40 m.p.h. twin-screw cruiser *Loraymo*. *H. Landshoff Photo*.

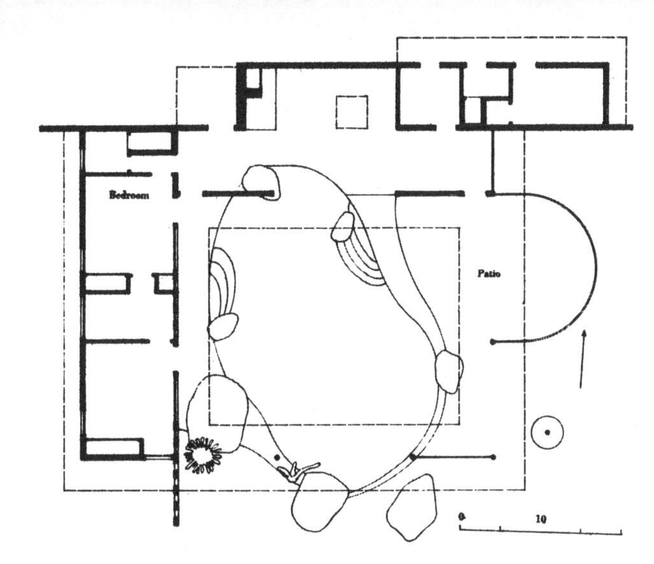

TIERRA CALIENTE: The plan of this desert house is compact and has proved most successful in operation. *Reprinted from House & Garden.*

View showing penetration of the pool into the house. Rectangle at right-hand corner is access hatch to underground pool-filtering and heating equipment. *Box-Slater Aerial Surveys.*

Living room, showing sliding glazed doors, open. Irregular shape on wall above commode is natural stone. Cabinet and wall section of pecky cypress. *Julius Shulman Photo.*

View taken from center of living room. *Bernard of Hollywood Photo.*

Bedrooms facing directly on the pool; steps lead into the water. Because desert nights are cold the spring water is heated to 80°. *Julius Shulman Photo.*

NEW YORK OFFICE. *Look Magazine Photo.*

Chapter 24

READER RIDES AGAIN

Industrial design as far as I am concerned is 25 per cent inspiration and 75 per cent transportation; and it makes Viola's life complicated. She seldom can plan a party, or a dinner, as I am likely to leave for California, Dayton, or London at a moment's notice. I have taken three round trips from New York to California in less than two weeks, flown to Europe across the Atlantic four times in four weeks. This flying business, on certain occasions, can become a first-class ordeal. It isn't necessarily bad weather, the food, boredom, or forced landings. More often the menace is small but vicious; it is the Crying Baby. Anyone who has flown eighteen hundred miles next to a little crying cherub complete with smell of sour milk and warm pipi will understand what I mean.

Last year while on a flight to Europe it was my lot to sit just across the aisle from such a little flying nuisance. He probably established a word record for yelling nonstop for seven hours at an average speed of 360 miles per hour. A tall, distinguished-looking old gentleman passenger, driven to the verge of nervous collapse, kept on exchanging panicky glances with me. When we reached that dreary refueling spot called Gander, my companion and I, limp and dizzy, had a few martinis and we considered the tactical situation. It was then that I noticed the airport nurse in the first-aid room in the process of filling the baby's bottle with warm milk. Our little passenger friend was going to be refueled too and made ready for its yelling leg to Paris. I considered the various medicine bottles on the shelf and it gave me an idea which I submitted to my companion. It was a gorgeous idea, he thought, and we should certainly try it at once. And to this day, there is an airport nurse, somewhere, who remembers the two passengers who tried to bribe her into mixing a tiny bit of chloroform with the baby's milk.

The passenger and I became fast friends and we have seen each other frequently since. His name turned out to be Henri Bernstein, France's greatest living playwright.

Anyone who travels is bound sometime or other to buy a chocolate bar, a pack of gum, or a nickel candy bar. My favorite is the candy bar. Not so much because of the product as of the container itself and its mystery. Those who by chance have read what is written on a candy package must admit that Americans can be candid and courageous to the extreme. They like to face the truth, regardless of how cruel it can be. A good example is the Pure Food Law, which compels a manufacturer to print on the wrapper the list of ingredients in the product. Take, for instance, an imaginary but representative package such as Driplets,

Jerklets, or Karamollo. It takes fortitude to buy and eat a product certified by the United States government as being

> made of sodium glutamate, dehydrated albumen, glycerine, carbohydrate derivatives, glucose, charcoal, lecithin, benzoate of soda, citric and oxalic acids, made with artificial flavor and certified U. S. color added.

Such chemical delicacies, sometimes pretty close to the formulas for TNT or paint removers, are consumed by the hundreds of millions annually. I often think about the brave research engineers who devote their lives to the development of a new unchartered type of candy. One must have a certain admiration for the candy kitchen here doing such inherently dangerous chemical experimentation in obscure anonymity, running the risk of being blown up at any moment by a new type of Honeychew or an unstable gumdrop. Anyone versed in chemistry shudders at what the explosion of an American marshmallow factory could do to a community. Such installations, like the Manhattan Project, ought to be restricted to desert areas, or else manufacturers should go back to old-fashioned candies that used to be made of sugar, maple syrup, cream, eggs, butter, nuts, etc. Nothing explosive.

But travelers themselves are running certain risks. Among the most current one is the common soda-fountain bacteria counter. As an industrial designer it has been my lot to spend some time on the other side of the counter in line of duty, while making an operational field study of a fountain. It would take Poe's talent in his most deliquescent vein to describe it, and Ivan Albright, the master of decay, to illustrate the story. Anyone who ever smelled the midsummer night stink of a sloppy soda fountain can never forget it. It is not a violent stink, it's a subtle blend of putrescence and semiputrescence, all half-tones. In general, the dominating scents are sour milk, decayed hamburger, chocolate, mustard, raspberry, onion, and vanilla. For anyone who has the pluck, I recommend one look at the jar in which the ice cream scoop soaks. It is hard to believe.

The customer's side of the counter has its subtleties too: coffee cups are lipstick-stained, china plates have thumb-prints, there is coffee in the sugar bowl and grease in your tumbler, and that well-known iridescent film of cold grease floats over the iced water. The sweat-drenched attendant is so busy that he can't afford more than a second to wipe off his brow on a corner of his chocolate- and tomato-stained apron before he again plunges his split thumb, Band-Aid and all, in *your* bowl of clam chowder. *Bon appétit* to you, dear reader.

Besides candy bar wrappers, I can recommend another kind of interesting literature for light reading. It is found on pamphlets enclosed within the packages of patent medicines. This grandiloquent prose usually starts as follows:

> The famous and world-wide properties of Zazapar, recognized for over fifty years as the classic treatment for hepatose, are in reality too familiar to require enumeration. Among its well-known properties are its action as an anti-toxi-desinterocolic vehicle for the arto-ami-benzo-steral contained in [etc., etc.].

Upon which it proceeds to enumerate in two thousand well-chosen words the well-known properties of Zazapar. It is cute reading for places where reading matter is scarce, such as bathrooms. The literature is usually available in the medicine cabinet of the average home.

Your telephone just rang, my friend Reader.

"Reader's residence."

"This is Long Distance; Raymond Loewy calling Mr. Reader from Los Angeles."

"Who?"

"Mr. Raymond Loewy, industrial designer."

"Just a moment, please."

You come to the phone.

"Yes, Loewy. How are you? Where are you?"

"Fine. I am in L. A., flying back to New York tonight for a couple of days. Say, last time you went on a trip with my boys, you said you'd like to go once more in about a month or so. You especially mentioned a trip connected with automobile design and also a West Coast job. I am planning just such a trip right now, with Rodney, one of our bright boys. We leave for South Bend Monday, planning to spend Tuesday and Wednesday at the Studebaker plant. Then on Wednesday afternoon we leave for Chicago, where we will have dinner with Franz, the manager of our office there. Then about midnight we fly to L. A. for a discussion of airliner interiors with the Transcraft people. It'll last a day or so. After that I plan to go back to my home in Palm Springs for a few days. Mrs. Loewy is there now. You might spend the weekend with us if you like."

"Sounds good. Sure, I'll come along. I want to know more about your work."

"Okay, make sure you take enough shirts and linen, as you won't have a chance to get anything laundered for five days. Better figure on two shirts a day. If possible, take one suitcase only and a small toilet kit. Rodney and I will meet you Monday on the "Century" in the center lounge one hour after departure.

Miss Bowman, our transportation girl, will send you your rail and air tickets tomorrow. See you Monday. Say, by the way, bring sunglasses along, cold tablets, and one extra suit. So long, glad you can come."

7 P.M. I wash my face in cold water, pick up *Time*, and walk to the center lounge where you already sit, having a drink.

"Nice to have you with us, Reader. The boys tell me that you had a good time together last time you went to Boston. Hope you won't be disappointed this time. Here's Rodney, coming over now. Rodney, meet my friend, Reader, who's going to accompany us on this trip. You know, same as he did recently with our Apex boys. Let's have a drink and tell him about our schedule."

I give you a condensed version of our South Bend visit. Roughly this is it: I was there ten days ago and we started three full-size mock-ups of a new front-end treatment for the 1951 models. The bodies are to remain as they are for another year. The front fenders also. The hood, radiator grille (or air-intake), hood ornament, fog lights, etc., are to be different. We are trying to get a new, more advanced appearance without losing the general flavor of the present production job, which has clicked with the public. Tomorrow we shall look at what our South Bend boys have developed since my last visit, and if everything is okay we will show the mock-ups to the management.

Rodney has had his second dry martini and he doesn't appear to be listening. His attention seems to be somewhere else. It is. He keeps on looking at a trim blonde in a gray tailored suit, two chairs down on the opposite side of the aisle. She reads *Esquire* absent-mindedly as she surveys our little group with one eye. Rodney acts a bit restless, especially every time she uncrosses and recrosses her nice legs in a fascinating technique. Being a modest young lady, she is careful that her gray skirt is chastely pulled down. The process, however, includes a well-timed split-second flesh exposure when her gloved hand lifts up the gray skirt high, quickly revealing a bank of pink thigh and lace before it is modestly pulled down again. Rodney is wildly interested and watches for the next shift with restless anticipation.

"Rodney," I say, "I was explaining our three mock-ups in South Bend. Which one do you favor?"

"Whom?"

"I say, which one do you favor?"

"No, thanks, I'll have one during dinner."

"I am not talking about martinis; I am talking about front ends."

"Oh."

"Aw, come on, let's have dinner," I say, as we leave. Rodney reluctantly follows after a last exchange of glances with the gray tailored kid.

Walsh, the "Century's" genial steward, welcomes us in the diner and we order. Rodney isn't hungry and keeps on watching who's coming into the diner. You select the Special 20th Century dinner (shrimp cocktail, steak, cherry pie, and coffee), and we start our meal, discussing tomorrow's problem.

Our tailored blond friend walks in and the steward directs her to the one vacant seat at a table for two. A gentleman is already sitting there, quietly dividing his attention between *Quick* and a glazed pork chop, 20th Century style. Within thirty seconds the gentleman passenger politely offers the lady a light. Fifteen minutes later they finish their second double martini. Rodney is dejected.

"Cheesis!" he says.

By the time we reach Poughkeepsie our couple is gay as they leave the diner. The train is fast and the curves sharp. Being a gentleman, the passenger politely steadies the lady along the center aisle, holding gently but firmly her trim waistline with both hands.

"Cheesis!" says Rodney.

Now that the interference has been removed, we can talk design. Rodney and I discuss a few points about tomorrow's mock-ups, making rough sketches on the back of the menu (which we later carefully tear up and throw away through a crack in the diaphragm between two cars. Designers must never leave sketches behind.)

This business of putting down ideas in rough-sketch form on any kind of available paper or scrap of paper may be surprising to the outsider and might appear to be a casual way to design products. All I can say is that I am an intense believer in the technique. The most imaginative work seldom starts at the drafting board. Some of the world's fundamental discoveries were made while shaving, during a walk, or riding in an elevator. I could mention dozens of outstanding commercial successes that originated in such a manner.

When a sketched idea appears interesting, we initial it, date it, and, upon arrival, turn it over to the technicians in our offices. They study it critically, develop it, draft it, and it is ready for further discussion.

"Why," you ask, "don't you carry small pads or sketchbooks instead of using sugar cube wrappers?"

We have tried it and it doesn't seem to work as well. Why? I don't know! Let our friend the psychologist isolate the motivating thought-mechanism behind it all. (Or should it be the psychiatrist?) Meanwhile we shall keep on sketching as we go, on whatever is handy.

A few years ago a Middle Western art institute wanted to have an exhibition of some of the very first sketches that we had made in the past—sketches that had become the starting point of many successful products or structures. When they saw the disreputable collection of old matchbooks, soda fountain menus, and torn envelopes we had to offer, they gave up the idea.

Elkhart. A biting wind slaps me in the face and wakes me up for good. The driver of Studebaker's courtesy car is waiting for us. Reader and I get the morning papers and we are on our way to South Bend, reading the *Tribune*. Rodney does not read newspapers. Those Russians again! They have stopped all traffic in and out of East Berlin for twelve hours—financial news—Li'l Abner, Dick Tracy, weather report (no good: Rain turning to freezing at night). Forty-minute ride among Indiana fields in the early morning light. Rodney yawns nonstop and gets us all yawning to tears. I can see the back of the chauffeur's head. Periodically his cheek-bones seem to widen out while his cap leans back a bit. He is yawning, too. One of these days, Rodney is going to make him doze at the wheel and we will all be killed in our sleep—in Indiana.

South Bend. We go to the Oliver Hotel for breakfast. It is only seven thirty in the morning but the Oliver's cafeteria is already filled. People get an early start in small Middle Western towns. Mabel, the attractive young waitress, greets us with a smile that warms up our hearts. She knows us well, as we are frequent visitors to the Oliver. If waitresses wore sweaters, she would make Jane Russell look skinny.

"Good morning, Rodney; good morning, Mr. Loewy." She nods at you questioningly.

"Oh, Mabel, this is Mr. Reader, a friend of ours from New York."

"Hi, Mr. Reader, glad to see ya."

The cafeteria is cozy, warm, and fragrant with coffee, waffles, and sausages. Outside, the February wind torments the black trees in front of the courthouse. Mabel is cute, we are cozy and full of good hot coffee; we feel grand, ready for the task. While waiting for the waffles, we discuss the problem ahead. I make some sketches on the back of our railroad ticket envelope. That's it! This is the solution to that instrument board. Rodney, not to be outdone, comes across with a smart idea for the tie-up between grilles and fenders. He sketches his idea briefly on the morning paper in the white margin alongside the Gumps and Little Orphan Annie. Here we are with two bright ideas already and we feel that the day starts well. Life is exciting. Rodney smiles at Mabel. Mabel smiles at Rodney.

We leave a "21" size tip to our friend, who is used to getting nickels, and we are on our way. Rodney stops at the cigar counter and buys two Baby Ruths, two Chicken Dinners, a bar of Hershey's, and a couple of Bit-O'-Honeys, which he starts devouring. The wrappers will come in handy in case we have more sketching to do.

Bob Bourke, the resident manager of the R. L. A. styling division at Studebaker, greets us when we arrive at our design department in Plant 3. It is a model of organization and modern design technology. It has been credited as having started new trends in body designing technique and it rates high in the automotive industry. It is well laid out, in a modern, air-conditioned building, with nearly a million dollars' worth of the latest equipment, specialized lighting, and machinery. Our forty men there are happy and efficient. Gracie, the receptionist, is a little beauty and it is a pleasure to see her again. You meet Gracie, who takes your overcoat and hangs it up in my office. It has a blue-gray carpet, comfortable chairs covered with black patent leather, gray walls, and plenty of golden light. A long shelf placed before gray drapes stretches all the way across the room; on it are dozens of colored sketches of hood ornaments, hardware, instrument panels, etc., ready for discussion. Bob and his assistant (also named Bob, which complicates matters a great deal) sit with us and we are about to review the morning schedule. They keep on watching you, then glance uneasily at me. Oh, yes, I forgot!

"Meet my friend, Mr. Reader, from New York. It is perfectly okay, Bob, you can talk freely in front of him, and let him see whatever we have. It has been cleared with the management and he can be trusted entirely. So let's proceed."

The two Bobs feel relieved and we resume our talk. Then we go through the department; look at everything, make suggestions and criticisms, review the situation in general. At twelve o'clock sharp, you, Rodney, the Bobs, and I drive to the executive cafeteria in the administration building, where the food is excellent. My friend H. V., the Chairman of the Board and President, is there with most of the top execs. It is a happy, cheerful group and I introduce you to many of them. Dick H., the head of the export company, and Paul C., the corporation lawyer, sit with us. Before long Rodney and I are making thumbnail sketches of tail-lights on the menu. This interlude gives you a chance to have a première Indiana showing of the snapshots of your wife and kids. It is a great success. Our friends pull out their own snaps, and after half an hour you know all about each other's family, average golf score, and Notre Dame's new halfback. Nobody discusses high taxes.

We drive back to "Styling" and resume work.

By evening, the mock-ups look fine and I call the management to set a time for our presentation tomorrow; then home to the Oliver for a bath, a short nap, and then dinner.

We meet at the bar with the two Bobs for a few drinks while Rodney gets more and more fascinated by the redhead who plays "Stardust" on a small organ. He likes to watch her cute feet playing with the slats and he is soon draped all over the organ, glass in hand, looking at her while she smiles and plays "Sweet Lorraine" and "You Go to My Head." Bob, Bob, and I have some ideas about interior trim for the cabriolet. Bob picks up a small easel-folded card on the table:

Now we have a couple of good sketches for the folding armrest in the rear compartment.

Time to drive to Alby's for a T-bone sirloin steak dinner. Rodney says he isn't hungry, but we finally manage to coax him away from the small organist, who plays "Good Night, Sweetheart" as we leave.

"Cheesis!" says Rodney. "With you it's always food, food, food."

Conversation during dinner is mostly about design, hot-rod races, or future style trends. Then to bed early, about ten o'clock. Rodney says he must stop at the drugstore, off the lobby, for a couple of aspirins; he says he is very tired and anxious to turn in. Good night, everybody.

An hour later I am in bed reading *Time* when I happen to think up an idea for a bumper guard that seems interesting. I should really take it up with Rodney right now so tomorrow we won't forget. So I pick up the phone and ask for Mr. Rodney Brandt now in the bar talking to the girl who plays the organ. In a few seconds Rodney is on the phone and I describe the bumper guard.

"Good night, Rodney, and take it easy."

"Oh, sure."

In the background I can hear the organ playing "Some Enchanted Evening." His headache is apparently cured.

Next day the meeting takes place. It is very successful. After the execs have left we make notes, make decisions, establish a schedule, and plan our next trip a week or ten days hence. Meanwhile, the company chauffeur who is to drive us to Chicago has bad news for us. The rain has stopped, but it is freezing and the highways are covered with ice. Driving one hundred miles over this stuff is just out of the question. We might not even reach Chicago by morning.

"Any trains?"

"No, next one is at 10:50 P.M. . . . too late for your air connection."

"What about the South Shore?" (A trolley that takes three hours.)

"Next one in fifteen minutes. You can make it if you hurry."

"Okay, thanks."

We wind up everything quick, rush out, and a Bob drives us to the trolley terminal as fast as the slippery roadway permits. We know now that driving to Chicago would be suicidal. Our yellow trolley is ready to leave; we rush in as the conductor shouts, "All aboard!"

The two cars are jammed with passengers on account of the bad driving weather and we stand for a while in the melted black sludge until someone gets off and we find seats—uncomfortable, narrow seats. What a dull trip, for three endless hours, with ten thousand stops.

We arrive in Chicago about 8:00 P.M. rather tired and we go to the Beachcomber for dinner. Franz, our Chicago manager, is there already and we discuss design problems during dinner between Moo Goo Gai Pan and Eggs Foo Young with plenty of rum punches. More sketches are made on cocktail lists, menus, and scraps of paper. We take it easy as the airport limousine doesn't leave the Palmer House until 1:00 A.M. About eleven Franz goes home and we wonder what to do to kill time.

Rodney says he vaguely knows a small night spot, the Oasis, or something, on Clark Street where they have a floor show. He doesn't know it very well he says, but we might take a chance. No cabs in sight. Oh, well, it isn't far, let's walk. So we start on the slippery sidewalk with our suitcases, leaning against the bitter Chicago night wind, looking for safe patches covered with sand or cinders. You, Reader, my friend, slip on the curb and fall on your side. You pick up your valise and we resume our safari in search of the Oasis. Your overcoat will need a good brushing. I had a fleeting notion that you looked a bit depressed. Oh, I must be mistaken.

Rodney looks across the street at a red neon palm tree, hesitates a moment, and says, "I think that's it."

We enter the hot and smoky joint while a small band plays a burlesque slow tune. Rodney's apparent hesitation about the identity of the place does not seem to jibe as he is immediately met by a girlie who wraps her arms around his neck and calls him "Poopsie, dear." Rodney acts a bit confused. Wednesday is a bad night at the Oasis, what with the weather and no conventions in town. A few plastered gents are scattered at small tables around the platforms on which Mlle. Paree does her patriotic strip with a blue-white-and-red G-string. Poopsie dear's girl friend sits at our table awaiting her turn. She wears a magenta rayon dressing gown and she looks like a quiet bourgeois young woman, most unsexy.

"Gee, it's swell to see you, Poopsie darling. I've had such trouble since you were here last month. Junie is over with the measles, but then Frankie got scratched by a cat and his arm got infected. I've had my hands full, I'm telling you; what with the sick kids, the laundry, the dishes, and this here job every night. Gee, that's nice of you, Rodney, to have helped me with that installment on the gas range. The dealer was going to reclaim it when I got your check. I'll repay you sometime, Rodney. Gee, I got an awful headache; I think I'm catching a cold. Say, whaddya think of these cold tablets? May I have a coke? What's your friend's name?"

It is all so familial and so proper. Poopsie dear apparently knew the Oasis pretty well after all. But she must leave us, for her turn is next. A moment later she appears on the platform and starts her strip while the band plays "Once in Awhile." Her figure doesn't add up to much. Somehow I can't disassociate her from Junie, Frankie, and the tiny apartment, the kitchen sink, the gas range, and the measles. Even Rodney looks thoughtful.

"Say, fellows, it's quarter past twelve, and we'd better get going."

So we are on our way again, suitcases and all, in the night. Fortunately, we find a cab. We arrive at the Palmer House; no airport coach in sight. Trouble?

"What's the matter?" we ask the porter.

"Flight's been canceled."

We call the airport. "That's right. Flight has been canceled on account of weather. However, there is a 3:00 A.M. flight on Atlantic that may go through. We can try to get you a seat if you wish."

"But there are three of us."

"I'll try. Can I reach you at some number?"

"Yes, we are in the lobby of the Palmer House. Just have me paged. Name is Raymond Loewy."

We hang around for a while in silent gloom, then the call comes through. Seats are okay. We are to take the airport coach leaving at 2 A.M. It is now 1:00 A.M. What shall we do? Too late to take a room. Too early for the airport. We wait in the lounge, now darkened. We are very tired and bored stiff. Rodney is yawning. I just do nothing, unable to sleep. About 2:00 A.M. the limousine arrives and we get on board. Only five customers, including a drunk one.

The ride through Cicero to Chicago's airport along Archer Avenue on a winter night is one of the world's dreary sights. For nearly an hour one sees nothing but the most dismal assemblage of blackened tenements, cheap stores, decrepit bars and hamburger joints; literally hundreds of them, all alike, all splashing an orange or blue neon sign in a streaky window, or above a sad door. The signs are nearly all beer signs. I am haunted by the inevitability of this beer sign parade along the way. The airport limousine is now dark, the drunken passenger is fast asleep. So is Rodney. I am just drowsy, watching the signs go by:

Schlitz
Pabst
BERGHOFF
and
Blatz

Schlitz
Pabst
BERGHOFF
and
Blatz

Schlitz
Pabst
BERGHOFF
and
Blatz

Tired, relaxed, bored, I begin to get hypnotized by the rhythmic obsession of the Cicero sonnet:

Schlitz
Pabst
BERGHOFF
and
Blatz

but a BLAZE OF WHITE FLOODS BREAKS THE SPELL:

ETHYL 22¢
TAX PAID

We stop for a red light, close to a trolleyful of charwomen, scarves tight around their heads, drowsing or reading the *Wharshawsky News*, or just staring blankly ahead. And the men, mostly old ones, miserable ones, the ones who have to take jobs at such hours, at that age, to keep alive. It looks more like a carload of

D. P.'s shipped to the salt mines than a trolleyful of American citizens. Poor wretches. Then we go on in our narrow canyon between rows of Blatzes and Schlitzes. Now and then we have variations:

PULATSKY POLISH SOCIAL CLUB

Fine Funeral Forty $

ZICHLER ON TAP

CHEESEBURGER

Matzlerbrau

and Schlitz

And suddenly a rash of bail shops, flashing off'n'on signs:

FINUCAN'S BAIL & BOND

WABASH BAIL

ARCHER'S BAIL

BOND-BOND-BOND-BAIL!

Then: Schlitz

Pabst
BERGHOFF
and
Blatz

Schlitz
Pabst
BERGHOFF
and
Blatz

Now we have passed the courthouse area and we resume the

Schlitz Blatz routine while our driver whistles, "Shine

on Harvest Moon." The passenger behind me must have the "lobster blues" as I have heard for the last ten minutes the maddening click, click of a quill toothpick in action. Rodney is fast asleep. You keep on looking at the black and white sign over the windshield.

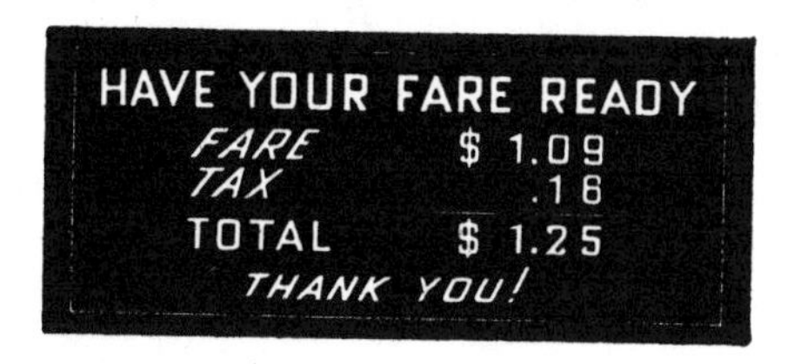

Schlitz - Pabst - BERGHOFF - Blatz

Parked right and left along the sidewalks are wretched-looking old sedans covered with impeccable snow that streamlines their angular shapes. Then a blue halo in the sky ahead of us. The AIRPORT, that glorious vision of tall, incandescent hangars, gleaming fuselages, beacons, and colored spots. It is alive and febrile, slick and virile. It is terribly exciting and American. Now, fully awakened, we rush to the sign,

Two or three passengers are there already, looking dejected. My traveling fifth sense perceives trouble ahead. Correct. Our flight is delayed.

"For how long?"

"About three hours."

We look at each other in utter disgust. We look around. What are we going to do? Take rooms and stay in Chicago? Impossible. Tomorrow's meeting in L. A. has been scheduled for a month. Engineers, airline executives, expect us for a final decision on the mock-up. We just have to be good sports about it and wait until flight time. Rodney says he knows a little bar across the highway, the Skylight Bar and Grill, where we can kill an hour. So we check our luggage and walk to the neon joint. A pretty cozy spot, smelling of fried onions and dollar perfume.

We sit at the bar next to a couple of airline stewardesses having a coke. They are going on duty and they don't look tired at all. Quite trim and fresh. They talk about a guy, Joe, who apparently is weatherman at the Kansas City Airport. He hasn't written the blond one for two weeks and things are sad. To Rodney this sounds interesting. Not too tired to sniff a possibility, he wakes up gradually.

He puts a quarter in the juke and we get Perry Como. I can feel that you, my friend Reader, are beginning to wonder how much you are learning about industrial design during this trip of ours. Believe me, you are learning all the fundamentals. Rodney is trying to get in on the gals' conversation, but it just doesn't work. Besides, he is pretty tired, and he does not insist. So we start talking about tomorrow's schedule, but we are not functioning well. We give up. We drag ourselves back to the terminal and just sit. And sit. Two and a half hours. Everything is so unattractive; it is all "cheap modern": yellow woods, mulberry fabrics, fluorescent lights, and orange trim. No trace of life but the moving sign on the automatic accident self-insurance machine. I have never taken one, but I do now, out of sheer boredom. I soon discover that it has entertainment value. As reading material alone, the policy is worth the investment:

FOR LOSS OF LIFE OR BOTH HANDS OR BOTH FEET OR
SIGHT OF BOTH EYES ....THE PRINCIPAL SUM
FOR LOSS OF ONE HAND AND ONE FOOT OR EITHER HAND
OR FOOT AND SIGHT OF ONE EYE..................THE PRINCIPAL SUM
FOR LOSS OF EITHER HAND OR FOOT OR SIGHT OF ONE
EYE...THE PRINCIPAL SUM
LOSS SHALL MEAN WITH REGARD TO HANDS AND FEET, ACTUAL SEVERANCE THROUGH OR ABOVE THE WRIST OR ANKLE JOINTS; WITH REGARD TO EYES, ENTIRE AND IRRECOVERABLE LOSS OF SIGHT. (IN SOUTH CAROLINA LOSS OF HANDS SHALL MEAN THE LOSS OF FOUR FINGERS, ENTIRE.)

Here is a nice problem to kill time: suppose, trying to be different to the end, that I lose two feet but no hand and lose only one eye. Do I get The Principal Sum? Or I lose only one hand but two feet and one eye? Or one eye, one foot, and two hands? What happens to The Principal Sum?

And here is a further complication:

EXCLUSIONS: THIS INSURANCE SHALL NOT COVER DEATH, DISMEMBERMENT OR LOSS OF SIGHT CAUSED WHOLLY OR PARTLY BY SUICIDE OR ANY ATTEMPT THEREAT (SANE OR INSANE) (IN MISSOURI WHILE SANE)

Mighty big of Mo!

Furthermore:

THE COMPANY SHALL HAVE THE RIGHT AND THE OPPORTUNITY TO MAKE AN AUTOPSY IN CASE OF DEATH

Sounds quite reasonable; I am not the one to deprive a guy of an opportunity.

My dismembered reverie is abruptly interrupted by the loud-speaker:

"Announcing the departure of Atlantic Airlines Flight Number 625 nonstop to Los Angeles at Gate Number 5. All aboard, please." Thank heavens! We get up stiffly, check in, walk up, pick a seat, buckle our belts, and get installed for the final stretch. Since 6:00 A.M. we have been on the go—twenty-four hours. I am so tired that I can't sleep. And tomorrow we have to be alert. I *must* sleep; just have to. I call the stewardess for a drink of water and I swallow a sleeping pill. Then, quiet. Peace at last. . . .

"What? What is that?"

"Mister! Mister!" she says.

"What? What is it?"

A gentle tap on my shoulder.

"Mister, please fasten your seat belt."

I don't understand very well. I have only slept such a short while.

"Are we in L. A. already?"

"No, we are going to land in Tulsa on account of weather."

"Where?"

"Tulsa, Oklahoma."

"Oh, no!"

The dread reality slowly reaches my sleepy brain. Things begin to materialize in their dreadful horror. Whatever morale was left in us "went to pieces all at once, all at once and nothing first." * Tulsa! We are completely out of our way. It's awful, really. Just too, too awful! Tulsa! We land, our plane taxies, comes to a stop with one last shudder, and everything is silent.

"Passengers will please carry all their belongings and wait in

* O. W. Holmes.

the air terminal until weather conditions improve. We shall keep you informed."

We get out in the cold, foggy dawn. The terminal is crowded; so many flights have been canceled. We finally discover three vacant seats. The unavoidable crying baby yells like a scalped coyote and gets everybody nervous. Besides, he has been air-sick and you can tell. I am sick, too. Airline-sick.

I go to the Atlantic Airline counter.

"I don't mean to ask any obscene question, mister, but how long do you think we may have to wait in this air palace?"

"We don't know yet, but we expect a weather bulletin any minute."

"What time is it now?"

"Seven forty-five."

We sit. Just sit. My sleeping pill is taking its full effect now and I am so tired. Wish I could stretch out somewhere. Then the loud-speaker again, in direct competition with the crying baby.

"Attention, please. Atlantic Airlines Westbound Flight Number 625 to Los Angeles is scheduled to depart at twelve forty-five noon."

Another five hours. Then we know what total dejection means.

You, friend Reader, ask us for three dollars in quarters, dimes, and nickels as you must make a call to New York, and we see you disappear into a phone booth.

"I am sorry, Mr. Reader," I say, "that we have had such bad luck. But in winter you have to expect those things, you know. Flying is still unreliable. Oh, but of course, Mr. Reader, I understand very well, and I am sorry to learn that your wife is sick, but you should be with her, really. I am glad we were able to get you a compartment on the train to New York. I know Mrs. Loewy will regret not to have you with us in Palm Springs for the weekend. Well, have a good trip. I hope Mrs. Reader will be well soon, and we shall see you in New York. So long."

You rush out and Rodney runs after you.

"Mr. Reader, Mr. Reader, here, don't forget your suitcase!"

"How stupid of me! Well, so long."

"So long."

Rodney and I look at each other.

"Cheesis," he says. "I wish I had a wife in New York, and she were sick, too."

Airlines are getting away with murder. I have flown steadily ever since the old Trimotored Fords went in service, back in 1927: my air mileage is well over a million miles. I understand, as any sensible passenger should, the difficulties of operating an airline system in winter when weather conditions are adverse. It is tough and deserves sympathetic understanding. It does not excuse stupidity and disregard of the passengers' feelings and comfort. In many cases, the attitude of the ground clerks borders on insolence. I have been lied to, misdirected, misinformed, duped, and waylaid at all hours of the day and night over the continental air map of the United States. I have been sent to the wrong airport at the right time, and to the right airport at the wrong time, told of canceled flights when the flights were on, and vice versa. I've been sent on three-hour bus trips in the middle of the night to find that the plane was leaving from the airport I just left. I've ridden back another three hours to be told that we had to go back where we came from, then to discover that there would be no flight at all until the next morning. Taking off at last, we were landed half-way through, in perfect weather, and told that the flight was terminated. Bad weather ahead? No, not at all. The plane was needed somewhere else. Experienced air travelers suspect that many such decisions making flying an ordeal are not always dictated by meteorological necessity. Shrewd airline operators may be tempted to get the maximum return out of the equipment and may interrupt a flight if there is a chance to run the ship on another schedule for better dollars-and-cents revenue. To the unfortunate passenger it means untold hardship that may reach near-exhaustion and leave him a nervous wreck. Sometimes a group of businessmen so abused and indirectly insulted may ask for an investigation of these schemes and tricks that are a disgrace to the industry. After all, airlines are public servants, and they have a responsibility. If the poor, abused railroads were trying to get away with that sort of stuff, Congress would be on their necks, ready for the kill.

It may be that airline ground attendants are underpaid. If so, give them a raise. Most customers would rather pay a bit more and be spared the trouble, the irritation, the fatigue, and especially the offhand treatment which is an insult to their intelligence.

Chapter 25

KEEPING FIT

As we see, the life of an active industrial designer is a strenuous one. Besides extensive traveling by train, plane, or motorcar in all kinds of weather, he must be alert and lucid upon arrival at his appointed meeting. It requires serious conditioning.

If the flying pest is the crying baby, the railroad has its own hazard. It is the plastered and talkative passenger who comes and sits at your table in the diner. Usually loud and vulgar, he attracts the attention of the other passengers to your table, where you would like to have your dinner in peace, read a magazine, or make business memos. Instead, you are unwillingly drawn into a cataract of incoherent talk, or long and pointless dirty stories. It is hard to escape, as the fellow would get offended and start making remarks about stuffed shirts, snooty guys, etc., etc. Other passengers, who have been on the same spot, feel sorry for you as you gulp your food, pay the bill, and rush out to safety.

There is another ordeal in store for the industrial designer. It is The Convention. Anyone who has had to attend a real one, a "giganic, mammoth jumbo" one in July in Cincinnati, you know what I mean.

In order to keep fit and function right, I have adopted a mode of living that alternates periods of intense work—often twelve hours or more a day, with constant traveling in all weathers—with periods of absolute rest. For instance, several weeks in the California desert, several months in Europe.

During the winter of 1945, I spent a week or so at the Racquet Club in Palm Springs; this was my first visit to the California desert and I fell in love with it. The coloring is lovely, the air crisp and pure; the nights are a dream. What I liked most was the subtlety of the coloring, the scents and the sounds; subdued, restrained, as if nature had blended everything into a delightful understatement. It is a symphony of delicate grays, beiges, horizon blues, and lavenders; the textures, whether harsh or delicate, are invariably exquisite. I felt very happy in Palm Springs, and I walked a great deal through the desert at the foot of Mt. San Jacinto. On one of these treks, I found an interesting spot on the side of a low hill, far away from other houses. It faced Indian territory (where no one is allowed to build) and it was a maze of granite boulders, some of gigantic dimensions, all pale gray in color. Part of a prehistoric glacier, they were magnificent with their pure white veins and striations. I thought it would be a perfect site for a small desert retreat. I went back twice, and became increasingly fascinated with the idea of building something there. One evening, after having had dinner at the club with my friends Charlie Farrell and Hoagy Carmichael, I made a rough sketch of what I had in mind on the Racquet Club's stationery. I went to bed, kept on sketching and by daybreak I was still working on plans and elevations. A small house already existed on paper. I slept a couple of hours and awakened again.

I shaved, took a cold shower, and went back to look at the site. It was perfect! I called on a local architect of whom I had heard, phoned a real estate man, and by that same evening I owned two acres in the desert. Next morning the architect and I went to the site with my sketches, and we located four very large and well-veined boulders. These would remain, and they would delineate the free form of the swimming pool. In fact, they would partially be submerged in its blue mountain water and form the nucleus of Tierra Caliente. With sticks and small stone mounds, we laid out the house right on the spot, so as to obtain the correct view and right orientation. During the next two days, I detailed the features and characteristics of both structure and landscaping; and when I flew back to New York, three days later, the bulldozers were already on the spot clearing the unwanted rocks away.

The house turned out to be a little gem. Compact, comfortable, and free from housekeeping complications, it is the perfect spot for relaxation. Viola and I are very happy there. It is one of the most cheerful, satisfying little houses I have seen. The pool, kept heated to eighty degrees, is not large, and it penetrates well inside of the living room, where the white carpet reaches its edge. Large sliding doors glazed with plate glass open up to thirty feet wide so as to blend inside and outside living. There are flowers in profusion, banana plants and palms both inside the house and outside. The transition between outdoor and indoor is nearly imperceptible. At night, a three-thousand-watt projector on top of a slender white-lacquered mast fifty feet high floods the forest of boulders with moonlight fluorescence. The pool is illuminated, and various palm trees, tall cacti, and octallias are bathed in golden light. When the house is kept in darkness, but for a log fire and candles, the sight is sheer beauty. A small fountain adds its frail tone to the silence of this oasis. In the distance, we hear the coyotes. Viola is near me; R. L. is happy.

At seven in the morning I wake up, put on a heavy robe over my pajamas, and drive to the Indian bath, a three-minute ride from Tierra Caliente. The Indian bath is a tiny shack, built right on top of a hot water spring, surging from the black desert sand in a corner of the Indian reservation. It is owned and operated by the Palm Springs Indians, and it is a primitive affair. You just go there, get into a tiny little room, and undress. Then you walk down two steps into a pool six by six and about two feet deep. The bottom is black sand kept under continual turbulence by the pressure of the surging hot water (130 degrees). You sit on the sand, alive with bubbles, as your body is massaged by the volcanic water. You stay about twelve minutes, getting progressively into a diluvian sweat. Then you go home. In my case, I quickly get dressed, all wrapped up in a flannel robe, a bath towel around my neck, and drive back home for a nap. When I wake up I take a swim in the pool, happy and relaxed. I feel wonderful. By that time it is 8:30 A.M. Pacific time and I transact my business by telephone with New York, Chicago, and South Bend, sometimes London or Paris. After lunch, usually taken at the edge of the pool, I start working on design or other problems until late afternoon.

We usually go to the gay Racquet Club for cocktails but we rarely lunch or dine out. Friends, either from the East or from Los Angeles, often come for cocktails or dinner. Jack Benny likes to spend a quiet hour with Viola and me, and we all relax and feel happy. Or it may be a movie director, or a starlet, or a star. William Powell, our neighbor, is often with us, with Mousie, his delightful little blond wife.

To us, Bill and Mousie are one of the main attractions of Palm Springs. Bill is a man of charm and wit, a brilliant conversationalist, and a good friend; he is a credit to his profession. While I was building the house, he would often visit with me and discuss various details. He was a bit concerned about the swimming pool in the living room. "Someday, some silly dope will fall in," he used to say.

When Tierra Caliente was completed and furnished, I asked a few friends to drop in for a little housewarming party. It was a warm desert night and a glamorous sight, with my decorative movie friends in gay summer clothes, toasting Tierra Caliente and their host with frappéd Roederer. Tony Martin came late, in a double-breasted blue serge, having just flown from a broadcast in Hollywood. About dinnertime, as my guests began to leave, I was at one end of the living room when Bill Powell waved at me. "Good night, Raymond, thanks and *à bientôt.*" I waved to him and Bill backed up—right into the water. He was there, rather bewildered, standing in the warm, shallow pool up to his waist with everybody laughing madly. Tony Martin, seeing his friend's embarrassment, did not hesitate a second and jumped in, blue serge and all. As a host, I felt it my duty to join them, so I calmly went down the four steps into the pool to join my wet guests, and the three of us shook hands very formally. My butler, a quick-witted fellow, spontaneously rose to the occasion: he placed a tray with three cups and a bottle of champagne at the edge of the pool, and Bill, Tony, and I drank one last toast to Tierra Caliente.

"Slim" Hawks (now Mrs. Leland Hayward) said to Hoagy Carmichael, "Raymond asked them to drop in, sometime. They certainly do!"

Viola and I spend the summer in France, part of it at la Cense, the rest on the Mediterranean, in the fisherman's village of Saint-Tropez. There we have a summer home situated at the water's edge on the side of a low cliff. Part of the house is built on columns directly over the ocean. The sea is clear, calm, and cobalt, flowers are plentiful, and our cook, Anna, is a master of southern French cooking. Her bouillabaisse is unsurpassed. She has been with me for twenty-five years, and two years ago her rather silly grandchild managed to have his left ear bitten off by an irritable ass. A slender bridge juts out thirty feet into the Mediterranean, and there we keep moored a fast and silent cruiser of American registry. (Can't be without a boat.)

It is a healthy life and we love the place, which I designed in 1925. We frequently cruise along the coast, to Eden Rock, Beaulieu, or Monte Carlo. Monaco's harbor is sumptuous and very popular with yachtsmen from all over the world.

In Saint-Tropez I have a chance to get in good physical shape, as we live in the sun and in the water most of the day. The *Loraymo II* is equipped with the necessary gear for deep-sea diving. Not to be confused with the familiar diving goggles or breathing tubes, this equipment enables the diver to reach relatively deep levels and to stay down an indefinite time. I have thus spent many unforgettable moments at the bottom of some sheltered cove, meandering in slow motion among the algae of the Mediterranean floor. It is rather disappointing to have to admit that nothing either hazardous or frightening ever happened to me during those explorations. One quickly gets accustomed to meeting octopuses and morays minding their own business. It is just delightful and I find the fishes friendly and unafraid. The only objectionable feature is the smell of hot oil in the air one breathes. The pump that supplies air to the diver through a rubber tube has a tendency to get hot in the summer sun. As it does, the lubricating oil gets hot too and the smell is quite nauseating. I have found that by mixing a few drops of Chanel 5 or some other good perfume with the oil, the situation is nearly corrected. Inhaling Stradivari while walking thirty feet underwater among the mackerels and octopuses is interesting. The diver, not being connected by phone to the surface, carries a few ping-pong balls with him. When he wishes to be hauled out, he merely releases a ball, which floats to the surface and signals his desire to his friends aboard the boat.

We usually spend the fall months, part of the winter, and the spring in New York. Our apartment is a small penthouse for which I made the layout. It is atop a building on Fifth-eighth Street that was erected in 1949. A great deal of planning went into the study of this compact residence, and it works well. Completely air-conditioned, it remains sealed all year long. Winter or summer, fresh, filtered air is pressurized and circulated inside, without ever opening a window. This eliminates all traces of dust and dirt. New York's bane, that sooty, gritty black dirt, never enters. The careful distribution of the air ducts gives one the impression of being in the open; the air is constantly in gentle motion, without trace of unpleasant "air-condition drafts."

I have never lived in a more pleasant city dwelling. The décor is a blend of classic furniture and good modern. Large wall areas are covered with acid-etched mirror, the color scheme is monochromatic, the textures varied and unusual. The Japanese style of floor contrasts has worked well. One steps from a high-glaze Tennessee marble floor in the foyer to a thick amethyst carpet laid on sponge rubber. The contrast is exciting. The penthouse is surrounded on three sides by a terrace fenced with wide bamboo stems bleached to a light beige. Two large Chinese lions of cerulean blue ceramic rest outside on bamboo bases, in front of two large windows. At night, Venetian blinds are left half closed. Two spotlights are focused on the blue lions, each casting its violent shadow on the bamboo fence. Seen through the horizontal blinds it is exotic and cheerful.

The general lighting is a subdued diffusion of warm light. Numerous lamps with opaque shades create the essential accents of highlights and shadows.

In each bedroom, a small bedside switchbox controls every light in the room, individually, and the electric blanket. There are many crystal chandeliers adding the warmth of candlelight to the coziness of the friendly living room. It is a spacious living room, where one feels relaxed and happy.

There we see our friends, a small group of delightful people who, thank heaven, are neither café society, bridge-club addicts, nor art faddists.

When partners come for dinner, it is always an occasion. We have great fun together—and it could never happen without the presence of our dear friends, "Honest John" Grant and "Mademoiselle" Reese.

Karl, our butler, and his wife Sally, a highly trained and decorative couple, take good care of us whether in New York or Palm Springs.

Thus we live a life which, to us, is ideal. It is a blend of everything that makes living interesting and eventful. America gives me the opportunity to be creative and imaginative. Europe—and France in particular—brings relaxation and perspective. This slowing down is imperative in order to retain a balanced outlook. It also gives me a chance to appreciate America even more keenly.

★

I am often asked to relate the interesting experiences I have had during my career as a designer. Here are a few I usually mention:

In order to design smoke-deflecting devices for high-speed steam locomotives, it is essential to chart exactly the flow of the airstream around a locomotive in action. A good method (empiric) is to observe the behavior of thin colored ribbons held at the end of a stick at different spots all over the locomotive at various speeds. This gives a clear indication of the behavior of the slipstream. Doing this while thundering along at a 90 m.p.h. clip, perched on top of the front coupling, or on the side catwalk, holding fast in the terrific hurricane is almost acrobatic. Or else standing up on top of the coal pile in the tender and passing under low bridges at nearly two miles a minute is quite a sensation. My love for speed mentioned earlier was at last requited.

Of a totally different character was the experience I had in the New York Supreme Court many years ago during a lawsuit over patent rights. A client of mine was suing another manufacturer for design infringement. It was a clear-cut case whereby the competitor had servilely copied the appearance of the successful product I had designed for my client. The defense argued that the product could not possibly be designed any other way and still function properly; therefore they argued the design patent was inoperative, or something to that effect.

The case had been dragging along for weeks without getting anywhere, until my client got fed up with the whole thing and asked me what to do about it. After a while, I had an idea and I suggested that he have his attorney call me as an expert witness. On the appointed day, I was called to the witness chair and I carried along a folding easel, a large drawing board, and some charcoal pencils.

The attorney for the plaintiff, my client, proceeded to ask me questions.

Q. "Mr. Loewy, what is your profession?"
A. "Industrial designer."
Q. "How long have you practiced your profession?"
A. "Since 1927."
Q. "Do you have your own business?"
A. "Yes."
Q. "How many designers do you employ?"
A. "About twenty."
Q. "How many clients do you serve?"

A. "About thirty-five."

Q. "Will you give us the names of some of them?"

A. "General Motors Corporation, Pennsylvania Railroad Company, International Harvester Company."

Q. "Have you examined the designs of the products as shown in Exhibits A and B?"

A. "Yes."

Q. "In your opinion could this particular product be designed in any other manner and still be practical and function properly?"

A. "Yes."

Q. "In your opinion, if you were commissioned to design this particular product could you create a design which would cause this product to look different from Exhibits A and B without affecting its practicability and functional qualities?"

A. "Yes."

Q. "Are you able to demonstrate such designs to the Court?"

A. "Yes, certainly."

Q. "How?"

A. "By making some sketches here and now."

By plaintiff's attorney to the Court:

Q. "Will the Court permit such designs and sketches to be made at this time?"

A. "It will; the witness may proceed."

I unfolded my easel, placed the drawing board on it, and started making rapid sketches in large black outline, visible to anyone in the back row. Ten minutes later, I had about twenty-five designs, all different, most of them attractive, all of them practical.

The attorney for the defense interrupted the proceedings and indicated to the judge the willingness of his clients to settle out of court. Within fifteen minutes the trial was wound up and the judge started on a new case.

I had another unusual experience in Washington. The Federal Government was conducting an investigation in reference to civil aviation. Having collaborated with Igor Sikorsky and the Greyhound Corporation in the development of an experimental twelve-passenger helicopter-bus, I was called as a technical witness. I went to the Federal Court Building carrying large drawings under my arm. I was directed to a certain court where I seemed to be expected, as an usher instructed me to unwrap my drawings and place them on a table directly in front of the judge's bench. His honor took one look at them, then at me, seemed completely dumfounded, and called an assistant. They all burst out laughing as it all became clear. I had been misdirected to a court where the United States post office department was trying *Esquire* magazine for improper use of the mail on account of certain risqué illustrations. The usher, expecting Varga, the *Esquire* artist, had mistakenly assumed that I was he.

In the design work we do for aircraft manufacturers my most exciting experience was not while flying but on land. While working at Lockheed, in Burbank (California), I was near the emergency escape hatch of a Constitution full-size mock-up. This door was a good sixty feet over the hard cement floor. One of the engineers opened the door. In front, at the top was a small handle exactly comparable to a strap hanger's handle in the subway. "This is a new escape device," he said. "The handle is connected to a steel cable, the cable to a spring-loaded reel. Should the wooden mock-up catch fire during construction, one just takes hold of the handle, jumps out, and holds on. It slowly decelerates the fall when it approaches the ground. Why don't you try it?" he said. I took one look at the concrete floor sixty feet below and went white. A couple of workmen were looking up at me kind of sarcastically, and my assistant, Harry Neafie, was in back of me. So there was nothing else to do but grab the handle and jump. Which I did. That free fall in the direction of the concrete is one of the reasons I have acquired silver-gray hair, standard equipment of the man of distinction.

During a design conference in my office in New York with a group of men from Wisconsin, I was interrupted by Helen Peters.

"Mr. Loewy," she said, "Mr. Dexter Brandon of the Department of State is on the phone. He has just arrived from Washington, he is at the Pennsylvania Station, and he would like to come and see you. What shall I say?"

"Did we make an appointment?"

"No, not that I know."

"Ask him, then."

Then Helen Peters again:

"No, Mr. Loewy, he did not have an appointment."

"Tell him I regret, but I am tied up and I will not have a minute today. Sorry."

Helen Peters explains the situation to the State Department caller and we resume our work.

A month or so later I was at my client's plant, having lunch in the cafeteria with the executives of the company. The president looked as if he had something on his mind.

Finally, "Mr. Loewy," he said, "my man told me about that call from the State Department you had during a meeting in your office. And how you would not see the gentleman in order not to interrupt the conference. Let me tell you that we understand and appreciate your attitude. But we want you to feel quite free with us. You see, here in the West, we are not as formal as you Easterners. We want you to feel at home with us. Next time you go right ahead, promise?"

"Promised."

I did not have the heart to tell him that Mr. Dexter Brandon was an electrician working in the State Department building as maintenance man. I happened to know him slightly and he wanted me to help him get a job with one of our clients in the appliance field. He had made quite a nuisance of himself by trying to rush things and calling at all times.

Will the reader excuse a little transition as I would like to insert here a very worth-while recipe. It is for a summer beverage that we often drink at our home in southern France.

CHAMPAGNE AND PEACHES:

Place a nice juicy peach previously peeled at the bottom of a tall glass. Half fill with cracked ice and add a jigger of Grand Marnier. Crush the peach slightly and fill up the glass with iced champagne. Drink while very cold.

A RECIPE FOR COFFEE CARAMEL SAUCE:

Take a pound of granulated sugar, one-quarter pound of butter, two pints of heavy cream. Place in a copper pan, blend well, and let it cook until it reaches the consistency of fudge. Add a tablespoonful of real vanilla extract and half a cup of very strong coffee. Let it simmer awhile. In order to test the consistency, pour a drop over a buttered plate and feel it with your fingers. To be right, it should be quite firm but not hard. Like chewy caramels. It should be served rather hot over good vanilla or coffee ice cream.

A RECIPE, AS YET UNPUBLISHED, FOR A SHERBET:

Prepare a mixture of one-third apricot nectar, one-third tangerine juice, and one-third pineapple juice. Add plenty of good champagne, a dash of fresh lime juice, and freeze in an ice-cream freezer. It is delicious.

Another simple recipe for anyone interested in a demonstration of complete happiness in this world of ours. I use it quite often and find it refreshing:

Take a good-size live dog fast asleep, preferably an Irish setter. Place gently and silently as close to its nostrils as possible a large chunk of liverwurst. Sit back and watch.

Stage One: At each intake of breath the scent of the sausage slowly permeates the unconscious "brain" of the subject until it reaches the boundaries of semiconsciousness. Then the nostrils begin to quiver slightly.

Stage Two: Lashes begin to flutter, saliva oozes out, and breathing evolves into sniffing.

Stage Three: Subject suddenly realizes the reality of the dream and in a violent convulsion lunges at morsel and swallows in one gulp.

Stage Four: This final stage is the most interesting one for the expert to watch, as it varies greatly according to individuals. Setters ordinarily express their utter bewilderment by sitting up and staring blankly, unable to decide what to do next. They remain there, unconvinced that such ecstasies exist outside of the world of dreams.

Some curious people are interested in finding out what constitutes the average day of a busy industrial designer's life. I point out to them that such a day must, as it does to all mortals, include two categories of events: those of a constant nature and the variable ones. In my particular case, the constants can be best described by going through a typical twenty-four-hour span in its chronological sequence.

The industrial designer (me), fast asleep, hears a gentle knock at the door. It is Karl, the valet-chauffeur.

"Good morning, sir, it is seven o'clock." He hands me the New York *Times* and a glass of orange juice. "It's very nice outside. A bit chilly." Newspaper—ten minutes of exercise that require superhuman determination. Shave, dress, breakfast; Sally, the maid, brings both the mail and some bit of news: "The day is quite nice but cold." The designer is ready to leave for his office. "Good-by, Sally." Elevator.

"Good morning, Mr. Loewy," says Ed, the morning elevator boy. "Kinda cold this morning." I make a vague statement expressing mild surprise and reconciliation with these ineluctable facts of meteorological life. Another Ed, the doorman, greets me and opens the door of the car with a rapid reference to the cold temperature that requires no further statement on my part. Karl drives on and meanwhile confirms his previous weather reports that the weather is chilly but quite nice.

We arrive at the office. Harry, the elevator starter, greets me with a cheery smile and his temperature estimation checks accurately with the situation already established and verified. During the ride up I am trying hard to remember something important I had wished to do immediately upon reaching my desk. Concentration is difficult because I am being briefed by Bob, the elevator operator, about the temperature characteristics of the morning. We reach the floor. Miss Marie Corbett, our lovely receptionist, greets me: "Good morning, Mr. Loewy." Nothing about the weather. She has been trained. I am safe at last. Safe until I must ride down the elevator again.

This is a simple morning. Sometimes I have to go to my dentist—or tailor—or patent lawyer, or all of them, before reaching the office. In such a case I try to condition myself to the weather-talk ordeal as well as I can. I haven't succeeded fully yet.

Another constant is the "coffee constant." I rarely drink coffee with my meals. This obvious defect has made my life terribly complicated. Especially in dining cars, where I take an average of six meals a week. Here is a typical dining-car meal cross section. I get in the diner, write my selection on the check, and add in block underlined letters NO COFFEE. The steward reads it and says immediately: "Will you have your coffee with your meal?"

"No, no coffee."

"Will you have milk or tea?"

"No thanks, no drinks."

The steward brings the first course:

"Will you have your coffee now?"

"No, thanks, no coffee."

"Would you like tea?"

"No, no tea."

Then he brings in the dessert and reopens the coffee matter.

"Would you like your coffee now?"

"No, I said no coffee. See here, it's on my check. See?"

He looks, "Milk?"

"No, I said no. I want no coffee, no tea, no milk, no chocolate. Nothing! Give me the check."

After a long wait, he returns with the check, and my cup of coffee.

Whenever I can, I stay in my office at lunchtime—this about three days a week. In this case, I send an office boy downstairs for a sandwich. The sandwich is always the same: Swiss cheese on toasted rye, with

NO BUTTER
NO LETTUCE
NO MAYONNAISE
NO MUSTARD
NO RELISH

After sixteen years in the same office, dealing with the same soda fountain, operated by the same owner, in our own building, I still cannot get my sandwich neat. After having my instructions repeated daily to both my secretary and the office boy, the result is practically certain: in 50 per cent of the cases, the sandwich includes:

BUTTER
or LETTUCE
or MAYONNAISE
or MUSTARD
or RELISH
or a combination of all five.

To the incredulous, may I give my assurance that this is not exaggerated. On occasions it has driven me to the verge of physical violence. When it comes on top of a morning where the weather has been abnormal (thereby creating furious conversation by elevator operators) it is almost too much for my nerves.

I am writing this aboard the 20th Century on my way to Elkhart. Today was an ordeal: (a) it has snowed; (b) I had lunch in my office with mayonnaise; and (c) I have just finished dinner (with coffee).

Chapter 26

WHERE TO?

As the reader must know by now, the author's opening promise that this book would be neither a treatise nor a philosophical essay has been pretty well kept. Without slipping into the dreadful maelstrom of amateur philosophy, there are nevertheless some clear signals for everyone to see and for me to state. For instance: as design consultant to more than one hundred of the nation's largest corporations, I am well aware of the staggering resources at the nation's disposal. Our scientific, technological, production, and population potential is so enormous that America has all it requires, not alone to raise its own standard of living, but also to help the rest of the world to do the same thing. To quote the great scientist Dr. Vannevar Bush, "The human enterprise—the endeavor of mankind from earliest days—is to force back barriers, no matter of what sort, in order that man's life may steadily grow to be a better life, physically, intellectually, spiritually." Enterprise seems to be a characteristic trait of the American man, and it has thrived under the free enterprise system. So has industrial design. If the system survives, industrial design can become a great factor in the struggle "to force back barriers," if not in the intellectual or spiritual theater, at least in the way of a better physical life. Private enterprise is America's destiny.

Extraordinary strides are being made in the field of pure science. American physicists, chemists, and biologists lead the world. With the help of the Rockefellers, Laskers, Sloanes, and Ketterings, cancer research is forging ahead and one can detect rays of hope, perhaps a solution within our lifetime. Industrially speaking, we make the rest of the world look like kids with a Meccano set. It certainly appears to be the century of the uncommon man: the American century. But how come that men so uncommon as a group can run the risk of becoming commonplace in their daily habits and manners? Why drift to such a standard of dullness? Must the fabric of our lives be an interweave of automatism and predigested thinking? Must it be dyed mouse-gray in apathetic surrender to bland routine? Can't we have our sandwich with mayonnaise—or without? And dinner *without* coffee?

What about the future of industrial design? What can it contribute to civilized life, and to American life in particular? How will it help us to reach the goal? In our pursuit of happiness, isn't peace of mind more important than any other type of bliss? Outside of good health, is there anything more precious? If not, then peace of mind is the true goal. The countless and incessant complexities and disturbances of everyday life are so many handicaps making this goal more difficult to reach. Sensory unpleasantnesses created by ugly form, color, or feel, noises, temperatures, or smoke are so many obstacles on the road to our destination. Faucets that leak, drawers that stick, hinges that sag, doors that warp are unnecessary irritations. Whether physiological or psychological, they can all be made less severe through correct planning.

They are all within the realm of the industrial designer. Transcending his early purpose, which was merely surface styling, the industrial designer becomes an integral part in the planning of every product, service, or structure. His presence at the inception stage will increase assurance that the end product shall be as free as possible of annoying features. Whether he deals with a bus seat, a carpet sweeper, or an airport terminal, the designer will try to make it more pleasant for you. He will be Knight of Good Taste and Defender of Your Nerves. He will see to it that all these material things that surround you remain in their proper place, unobtrusive and silent. If he succeeds, it will remove much turbulence and complication from your life. For I believe that the process through which one reaches intellectual and spiritual peace of mind can take place in physical comfort just as well as in constant petty annoyance.

The ascetic who lives in solitude may scoff at this sybaritic approach, but what is he doing himself? Just this: he is escaping civilization's interference with his concentration work. If peace of mind is the reward of a life of love, abnegation, self-denial, and devotion, I don't see why all this can't be achieved without zippers that jam, hinges that creak, and pens that leak. Anyone who welcomes such discomforts as necessary parts of his own special and private Calvary shows a strange lack of imagination in the choice of his unpleasantnesses. I could mention a set of others that would generate at least as much distress in a vastly more interesting way. I am all for more unusual Calvaries.

I believe that the first example of mass production in man's history was the invention, in the tenth century, of the mechanical clock. The units thus mass-produced were seconds and minutes. Man began to think of minutes and hours as material that belonged to him alone. They were his property. So he began the process of systematic allotment of this time-material for the best results to his welfare: so much for sleep, so much for work, so much for leisure. It probably marked the beginning of planned effort and planned leisure, the start of organized living. This fundamental move was far more important than accessory inventions such as the steam engine or even the wheel. So man abandoned the system of physical time-impulses such as day and night, and adopted the regulated mechanized time-units of seconds, minutes, and hours. Efficiency gained and liberty lost a great deal.

Ever since prehistoric days, man has worked in daytime and slept at night; he did not work at night because he could not see. He worked more in summer than in winter, as the daylight lasted more. When too tired, he would stop working altogether. A recurrent rest period was felt to be a necessity; he worked for six days, rested on the seventh, and the Week was born. The "six-day work, one-day rest" cycle was then best adapted to the physical efforts involved.

But now conditions are different. With excellent artificial lighting and air-conditioning (or climatizing), work at night is just as easy as, or easier than, in daytime, the level of brightness being kept at a constant. In fact, many of the latest plants, stores, and offices are already daylight-sealed and work is done under manufactured illumination. With the coming availability of unlimited energy through atomic power, the climatizing of whole plants, even communities, perhaps entire regions, may become a fact. Coupled with this, improved equipment and technics will require less physical effort and far less time for a given output.

So it would seem that the present "daytime-for-work, night-time-for-sleep" cycle may not be the best basis for future living. And it may be that the present "week" has ceased to be the right answer. One can see a possibility of daytime becoming leisure time (more outdoor living) and night being assigned to both work and sleep. For instance:

WORK:	Midnight to 8 A.M.
LEISURE:	8 A.M. to 4 P.M.
SLEEP:	4 P.M. to midnight

As far as the week is concerned, it might become this:

WORK:	9 days
REST:	5 days

The possible advantages would be these: on account of the certainty of cheaper, faster transportation, the worker would be able to "get away" for a long period, the equivalent of a vacation; could go to far places and enjoy things and climates unknown to him at present; could build up his health and the health of his family. As far as work is concerned, owing to the improved manufacturing setup, nine days would not be too much of a physical strain.

There are naturally plenty of adverse factors, but in general, the conception seems to justify further exploration. I realize that a main objection might be offered on religious grounds, as the position of Sunday would become a precarious one. I can assure my religious readers that I fully understand this viewpoint, that I respect it, and that the above suggestion is not meant to offend anyone's feelings. Besides, the author has given fair warning that philosophy is not exactly his dish and he feels ready to drop the whole thing at the slightest provocation.

But, just the same, redesigning the week would be loads of fun and it would certainly confirm the fact that industrial design covers a lot of ground. Next job might be the redesigning of the G.O.P. There is nothing too tough for us. And let's remember what Montaigne said: "No one is exempt from talking nonsense; the misfortune is to do it solemnly."

The other night I saw a short movie newsreel that I will never forget. Unexpectedly sandwiched between a basketball game and ice-fishing on Lake Michigan, the film put me in a state of emotional shock. I felt as if I were witnessing the birth of something even more staggering than atomic energy. The impressive part of it is that it was such a gentle, orderly phenomenon, one might almost say a peaceful performance. It was simply this: an automatic movie camera had been rigged up alongside one of the Army's new experimental rockets, and it recorded its stratospheric flight up to its apex thirty-seven vertical miles away. Then, the camera parachuted down to earth, intact with its precious record.

I imagined myself being in such a space ship as it started very gently from its launching site in New Mexico. One could see the rocky desert slowly fading away as the rocket gathered speed. Several miles up, the mountains became molehills, and as our sky-vehicle soared up smoothly and silently, I began to relax as one would in the comfortable seat of an airliner. By now the mountains were nearly flattened out and very pale; large rivers were thin black threads and the earth seemed to acquire a lunar quality. At the horizon, the planet's outline became a sharp curve and the earth gradually took the appearance of a section of a globe. As we silently climbed higher still, the sensation of earthly detachment became very real and overwhelmingly pleasant. It was something I never knew could exist. It was all so smooth, so uneventful. We were simply leaving the earth.

Then we began to decelerate, our sky-home lost its speed, the camera lost its direction, the screen went blank, the newsreel shifted to a new subject: we were right in the middle of a Ku Klux Klan Konklave somewhere in Alabama.

I never felt more frustrated in my life.

So I had seen it happen over again. After the light bulb, the telephone, the phonograph, the cinema, aviation, radio, television, now the beginning of interplanetary transportation! It felt so near, so plausible.

Should the human race decide not to annihilate itself for some cause—most righteous, no doubt—the second half of this century should be a fairly exciting one to live in.

I like the story of the boy scout reporting the good deed of the day to his master:

"And what have you done, Ray?"

"Walter, Henry and I helped a lady across the street."

"Very nice. But why did it take the three of you?"

"The old lady did not want to cross."

September, 1949, marked the thirtieth anniversary of my landing in America. What is the score? Have I done my good deed? Or am I one of the three boy scouts? Maybe America did not wish to cross Style Street, after all. It is not for me to answer or to judge. Meanwhile, I am going to try harder than ever because I think I ought to and because I like it here and would like to help improve it further. For designwise, and regardless of its shortcomings, the American Way of Life is at present the most advanced model on the planet. And I just happen to like advanced models.

But that isn't enough. A recent survey established the fact that products manufactured after R. L. A. design specifications reach three billion dollars a year. There is attached to this sum not alone the trust put in us by American industry; there is a social responsibility of which my partners and myself are acutely conscious.

It is for us to devote all our effort, our intelligence, and our talents toward one vital goal: the lowering of the cost of manufactured goods. Thus can we both speed up employment and bring more essential products to the underprivileged classes, a major segment of our population which needs, more than any other, the labor-saving equipment or services that will lighten its burden. This is democracy in action. America has treated me with such decency, such fairness, that I will try to do my humble part well, in appreciation and gratitude. As Samuel Gompers once said, "The greatest harm that management can do to labor is to operate at a loss." I will see to it that our clients are kept out of the red.

So long, dear reader. Hope to meet you soon in one of America's latest satellite resorts, perhaps having luncheon (shrimp cocktail, creamed chicken with peas and carrots, ice cream and coffee) around the new-type dehydrated gravity swimming pool. (Probably designed by Raymond Loewy Associates.)

ABOUT THE AUTHOR

RAYMOND LOEWY *first dreamed of building cars and locomotives in Paris, where he was born in 1893 and where he spent the first twenty-six years of his life. The third of three sons, Raymond filled his school notebooks with so many sketches of locomotives, automobiles, and airplanes that his parents sent him to engineering school. But at twenty-one, the student engineer was called off to World War I as a private. After four years of war, he came out a captain decorated with the cross of the Legion of Honor, the* Croix de guèrre, *and four citations. He came to America and went to work as a fashion illustrator for* Vogue *magazine. In 1927, after redesigning an old-fashioned duplicating machine, almost overnight fashion artist Loewy decided to become an industrial designer. His big chance came in 1934, when he was commissioned to dress up a refrigerator, and with the phenomenal success of Mr. Loewy's new functional refrigerating unit, his reputation was made.*

Mr. Loewy spends part of the winter working in the dream house he designed and built in the desert near Palm Springs, California. Summers always find him back in France, where he has three homes: a red-tiled villa overlooking the Côte d'Azur at Saint-Tropez, a sixteenth-century manor outside Paris, and a Paris apartment. And now, as head of the largest industrial design organization in the world, Raymond Loewy has written his own story, The Personal Record of an Industrial Designer from Lipsticks to Locomotives.